Remote Sensing Image Processing

Synthesis Lectures on Image, Video, and Multimedia Processing

Editor
Alan C. Bovik, *University of Texas, Austin*

The Lectures on Image, Video and Multimedia Processing are intended to provide a unique and groundbreaking forum for the world's experts in the field to express their knowledge in unique and effective ways. It is our intention that the Series will contain Lectures of basic, intermediate, and advanced material depending on the topical matter and the authors' level of discourse. It is also intended that these Lectures depart from the usual dry textbook format and instead give the author the opportunity to speak more directly to the reader, and to unfold the subject matter from a more personal point of view. The success of this candid approach to technical writing will rest on our selection of exceptionally distinguished authors, who have been chosen for their noteworthy leadership in developing new ideas in image, video, and multimedia processing research, development, and education.

In terms of the subject matter for the series, there are few limitations that we will impose other than the Lectures be related to aspects of the imaging sciences that are relevant to furthering our understanding of the processes by which images, videos, and multimedia signals are formed, processed for various tasks, and perceived by human viewers. These categories are naturally quite broad, for two reasons: First, measuring, processing, and understanding perceptual signals involves broad categories of scientific inquiry, including optics, surface physics, visual psychophysics and neurophysiology, information theory, computer graphics, display and printing technology, artificial intelligence, neural networks, harmonic analysis, and so on. Secondly, the domain of application of these methods is limited only by the number of branches of science, engineering, and industry that utilize audio, visual, and other perceptual signals to convey information. We anticipate that the Lectures in this series will dramatically influence future thought on these subjects as the Twenty-First Century unfolds.

Remote Sensing Image Processing
Gustavo Camps-Valls, Devis Tuia, Luis Gómez-Chova, Sandra Jiménez, and Jesús Malo 2011

The Structure and Properties of Color Spaces and the Representation of Color Images
Eric Dubois
2009

Biomedical Image Analysis: Segmentation
Scott T. Acton and Nilanjan Ray
2009

Joint Source-Channel Video Transmission
Fan Zhai and Aggelos Katsaggelos
2007

Super Resolution of Images and Video
Aggelos K. Katsaggelos, Rafael Molina, and Javier Mateos
2007

Tensor Voting: A Perceptual Organization Approach to Computer Vision and Machine Learning
Philippos Mordohai and Gérard Medioni
2006

Light Field Sampling
Cha Zhang and Tsuhan Chen
2006

Remote Sensing Image Processing

Gustavo Camps-Valls, Devis Tuia, Luis Gómez-Chova, Sandra Jiménez, and Jesús Malo

ISBN: 978-3-031-01119-1   paperback
ISBN: 978-3-031-02247-0   ebook

DOI 10.1007/978-3-031-02247-0

A Publication in the Springer series
SYNTHESIS LECTURES ON IMAGE, VIDEO, AND MULTIMEDIA PROCESSING

Lecture #12
Series Editor: Alan C. Bovik, *University of Texas, Austin*
Series ISSN
Synthesis Lectures on Image, Video, and Multimedia Processing

Print 1559-8136   Electronic 1559-8144

Remote Sensing Image Processing

Gustavo Camps-Valls, Devis Tuia, Luis Gómez-Chova, Sandra Jiménez, and Jesús Malo
Universitat de València, Spain

SYNTHESIS LECTURES ON IMAGE, VIDEO, AND MULTIMEDIA PROCESSING #12

ABSTRACT

Earth observation is the field of science concerned with the problem of monitoring and modeling the processes on the Earth surface and their interaction with the atmosphere. The Earth is continuously monitored with advanced optical and radar sensors. The images are analyzed and processed to deliver useful products to individual users, agencies and public administrations. To deal with these problems, remote sensing image processing is nowadays a mature research area, and the techniques developed in the field allow many real-life applications with great societal value. For instance, urban monitoring, fire detection or flood prediction can have a great impact on economical and environmental issues. To attain such objectives, the remote sensing community has turned into a multidisciplinary field of science that embraces physics, signal theory, computer science, electronics and communications. From a machine learning and signal/image processing point of view, all the applications are tackled under specific formalisms, such as classification and clustering, regression and function approximation, data coding, restoration and enhancement, source unmixing, data fusion or feature selection and extraction. This book covers some of the fields in a comprehensive way.

KEYWORDS

remote sensing, Earth observation, spectroscopy, spectral signature, image statistics, computer vision, statistical learning, vision science, machine learning, feature selection and extraction, morphology, classification, pattern recognition, segmentation, regression, retrieval, biophysical parameter, unmixing, manifold learning

Contents

Preface

Remote sensing of the Earth from space is changing our lives continuously. The weather forecasts now look impressively accurate! Agriculture also benefits from a more accurate monitoring of the Earth's processes, and from a much closer look at the phenological periods, which allow farmers to improve their harvests. And what about the oceans? With new satellite sensors, we can now measure the salinity of the oceans and estimate their temperatures very precisely. All of this was unthinkable fifty years ago. Nowadays you may get all this information with a few clicks on your computer or smartphone.

Many satellites with several onboard sensors are currently flying over our heads and many are being built or planned for the next years. Each one has its own specificities, pushing further the boundaries of resolution in spatial, spectral or temporal terms. However, these advances imply an increased complexity, since the statistical characterization of remote sensing images turns out to be more difficult than in grayscale natural images. New problems must be faced, like the higher dimensionality of the pixels, the specific noise and uncertainty sources, the high spatial and spectral redundancy, and the inherently non-linear nature of the data structures. To make it even better, all these problems must be solved in different ways depending on the sensor and the acquisition process.

On the bright side, what looks like problems for the users also constitute research challenges with high potential for the communities involved in the processing of such information: the acquired signals have to be processed rapidly, transmitted, further corrected from different distortions, eventually compressed, and ultimately analyzed to extract valuable information. To do all of this better and better, the field of remote sensing research has become multidisciplinary. The traditional and physics-based concepts are now complemented with signal and image processing concepts, and this convergence creates a new field that is capable of managing the interface between the signal acquired and the physics of the surface of the Earth. Add to these computer vision and machine learning, and you get a whole lot of new exciting possibilities to understand and interpret the complex and extremely numerous images that satellites provide everyday.

Obviously, this is not the first book on remote sensing. There are many excellent treatises on this topic, but most of them are either specific to the application (land, atmosphere or oceans) or general-purpose textbooks (dealing with the physics of the involved processes). There are surprisingly few books dealing with the image processing part of the field and, due to the rapid evolution of the field, they are often outdated. This volume aims at filling this gap, by describing the main steps of the remote sensing information processing chain and building bridges between the remote sensing community and those who could impact it by sharing their own view of the image processing problem.

The book focuses on what we consider the most active, attractive and distinctive aspects of the field. It is organized in six chapters, which relate to other scientific communities (see Fig. 1). Chapter 1 offers a short review of the underlying physics, a list of the major applications, and a survey of the existing satellite platforms and sensors. We will present what remote sensing images look like, as well as the main preprocessing steps prior to proper image analysis. This chapter also provides pointers to companies, publication journals, conferences and main institutions involved in the field. Chapter 2, studies the statistics of hyperspectral images and compares them with those of natural gray-scale and color images. Some interesting conclusions are established: while spectral and spatial features certainly differ, the main statistical facts (spatial and spectral smoothness) are shared. This observation opens the door to new researchers from the classical image processing and computer vision communities. Chapter 3 deals with the problem of feature selection and extraction from remote sensing images: apart from the standard tools available in image/signal processing, we will see that specific physically-based features exist and need to be used to improve the representation, along with new methods specially designed to account for specific properties of the scenes. Chapter 4 tackles the important and very active fields of remote sensing image classification and target detection. Problems with describing the land use or detecting changes in multitemporal sequences are studied through segmentation, detection and adaptation methods. Researchers in this field have come up with many efficient algorithms, and these fields are perhaps the most influenced by the machine learning community. Chapter 5 presents the field of spectral unmixing analysis, which studies ways to retrieve the different materials contained in spatially coarse and spectrally mixed pixels. This relatively recent field shows interesting connections with blind source separation, sparse models, geometry, and manifold learning. Chapter 6 addresses one of the most active field in current remote sensing data analysis: the area of model inversion and parameter retrieval, which deals with the retrieval of biophysical quantities such as ocean salinity or chlorophyll content. This field is related to regression and function approximation, search in databases, regularization, and sparse learning.

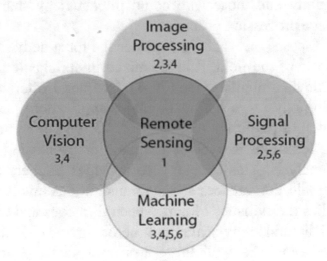

Figure 1: Roadmap: The four main fields involved in remote sensing image processing and the most related chapters of this book.

After reading just these few pages, you can see that the field of remote sensing image processing is indeed very broad. For this reason, we could not cover all the interesting and relevant aspects exhaustively. For example, we do not cover radar or LiDAR signal processing, which are very active areas of research. Also, while we pay attention to the image features, the key issue of noise sources and corrections is not treated in detail. The same is true for the fields of image compression, super-resolution, co-registration, deconvolution or restoration to name a few. We preferred to be consistent with our presentation and leave room for a second volume!

The book is mainly targeted at researchers and graduate students in the fields of signal/image processing related to any aspect of color imaging, physics, computer vision and machine learning. However, we hope it will also find readers from other disciplines of science, and in particular in the fields of application where remote sensing images are used.

Gustavo Camps-Valls, Devis Tuia, Luis Gómez-Chova, Sandra Jiménez, and Jesús Malo València, 2011

Acknowledgments

This book would have never been possible without data and code kindly shared by some colleagues. We would like to thank Dr. Allan Nielsen from Denmark Technical University for sharing some feature extraction code at his web page, Dr. J. Bioucas-Dias from the Instituto Superior Técnico in Lisboa for his unmixing tools, Dr. Chih-Jen Lin from the National Taiwan University for his wonderful libSVM and Dr. Jordi Muñoz form the University of València for tuning libSVM in order to allow us to get exactly what we needed. We also would like to thank Dr. Carl Rasmussen from the University of Cambridge for his great Gaussian Process toolbox and Dr. M. Tipping now at Vector Anomaly for his implementation of the relevance vector machine.

A book about image analysis would be quite dull without images. For these, we would like to thank the following persons who kindly provided the datasets used in this book: Prof. Paolo Gamba, University of Pavia (Pavia DAIS data), Prof. M. Kanevski, University of Lausanne (Brutisellen QuickBird data), Prof. D. Landgrebe, Purdue University (Indian Pines), Prof. Melba Crawford, Purdue University, for the KSC AVIRIS data, Dr. Xavier Calbet, EUMETSAT, for the provided IASI data and experiments with the OSS radiative transfer code, Dr. José F. Moreno, University of València for kindly providing access to data from the SPARC campaign, the European Space Agency for providing MERIS images acquired in the frame of the "MERIS/AATSR Synergy Algorithms for Cloud Screening, Aerosol Retrieval, and Atmospheric Correction", J. Z. Rodríguez, Valencian Institute of Cartography for the citrus groves images, and NASA and the USGS team for sharing the Cuprite dataset and the corresponding spectral library.

Last but not least, we wish to thank colleagues who provided constructive and comprehensive reviews for this book: Dr. Luis Guanter from Oxford University, Dr. Raúl Zurita-Milla at the University of Twente (ITC), and Drs. Javier García-Haro and Jochem Verrelst at the University of València. Special thanks go to some collaborators through the years whose efforts are somehow reflected in this lecture book: Dr. Frederic Ratle for his collaboration on semisupervised neural networks, Dr. Jerónimo Arenas-García for results on kernel feature extraction, Dr. Jochem Verrelst for results and discussions on the application of GP regression, and Emma Izquierdo and Julia Amorós for their excellent work on citrus grove identification and image photointerpretation.

Finally, we would like to acknowledge the help of all the people involved in the collation and review of the book. Without their support the project could not have been realized. A special thanks goes also to all the staff at Morgan & Claypool publishers, and in particular to Joel Claypool, who worked hard to keep the project on schedule and to Prof. Al Bovik, who initiated this adventure and kindly invited us to contribute to this series.

Gustavo Camps-Valls, Devis Tuia, Luis Gómez-Chova, Sandra Jiménez, and Jesús Malo València, 2011

CHAPTER 1

Remote Sensing from Earth Observation Satellites

This introductory chapter summarizes the field of remote sensing for Earth Observation (EO). The chapter provides the most important definitions for a beginner in the field, and illustrates how remote sensing images look like before and after pre-processing. In addition, we list the main applications and review the existing satellite platforms and sensors. Finally, the chapter contains pointers to companies, publication journals, conferences and main institutions involved in the field.

1.1 INTRODUCTION

1.1.1 EARTH OBSERVATION, SPECTROSCOPY AND REMOTE SENSING

The main general goals of Earth Observation through remote sensing can be summarized as follows:

1. *"Monitoring and modeling the processes on the Earth surface and their interaction with the atmosphere."*

2. *"Obtaining quantitative measurements and estimations of geo-bio-physical variables."*

3. *"Identifying materials on the land cover analyzing the acquired spectral signal by satellite/airborne sensors."*

To attain such objectives, the remote sensing community has evolved to a multi-disciplinary field of science that embraces physics, biology, chemistry, signal theory, computer sciencc, electronics, and communications. These objectives are possible because materials in a scene reflect, absorb, and emit electromagnetic radiation in a different way depending of their molecular composition and shape. Remote sensing exploits this physical fact and deals with the acquisition of information about a scene (or specific object) at a short, medium or long distance. The definition of the distance is arbitrary: one might think of photographic cameras as devices for remote sensing, as well as hyperspectral cameras or radar antennas mounted on satellite sensors. While this might be just a question of terminology, we argue here that the distinction should come merely from statistics. In this sense, the next chapter will analyze the statistics of satellite hyperspectral images and will compare it with that of natural photographic images.

1.1.2 TYPES OF REMOTE SENSING INSTRUMENTS

Attending to the type of *energy sources* involved in the data acquisition, two main kinds of remote sensing imaging instruments can be distinguished:

- *Passive* optical remote sensing relies on solar radiation as an illumination source. The signal measured at the satellite by an imaging spectrometer is the emergent radiation from the Earth-atmosphere system in the observation direction. The radiation acquired by an (airborne or satellite) sensor is measured at different wavelengths and the resulting spectral signature (or *spectrum*) is used to identify a given material. The field of *spectroscopy* is concerned with the measurement, analysis, and interpretation of such spectra [Danson and Plummer, 1995, Liang, 2004, Lillesand et al., 2008, Richards and Jia, 1999, Ustin, 2004]. Some examples of passive sensors are infrared, charge-coupled devices, radiometers, or multi and hyperspectral sensors [Shaw and Manolakis, 2002].

- On the other hand, in *active* remote sensing, the energy is emitted by an antenna towards the Earth's surface and the energy scattered back to the satellite is measured [Mott, 2007, Wang, 2008]. Radar systems, such as Real Aperture RAR (RAR) or Synthetic Aperture Radar (SAR), are examples of systems for active remote sensing. In these systems, the time delay between emission and return is measured to establish the location, height, speed, and direction of objects.

In this book, we will only focus on passive sensors since the product delivered from the acquisition, unlike radar signals, can be better interpreted as an *image* in the sense of natural images. The field is also very interesting since have experienced a great evolution in the last decades in terms of the quality of the acquired images (both in spatial, spectral, and temporal resolutions). This ever-growing evolution increases the difficulty of the signal/image processing problems and the need for improved processing tools. This may be certainly very motivating for the machine learning and signal processing communities.

In the case of optical (passive) remote sensing, one could alternatively classify the field into three main categories by looking at the *wavelength regions* of the spectrum: (1) visible and reflective infrared remote sensing; (2) thermal infrared remote sensing; and (3) microwave remote sensing. Throughout this book, we will show examples of remote sensing processing of images covering all these regions.

1.1.3 APPLICATIONS OF REMOTE SENSING

Analysis of the acquired multi-dimensional images allows us to develop real-life applications with high social impact, such as urban growing monitoring, crop fields identification, disaster prevention, target detection, or biophysical parameter estimation. Remote sensing makes it possible to collect data on dangerous or inaccessible areas, and to monitor Amazonian deforestation, measuring the effects of climate change on Arctic and Antarctic regions, and sounding of coastal and ocean depths. Remote sensing also replaces costly and slow data collection on the ground, ensuring in the process that areas or objects are not disturbed. Orbital platforms collect and transmit data from different parts of the electromagnetic spectrum, providing researchers with useful information to monitor cities, detect changes, and to efficiently manage natural resources such as land usage and conservation Liang [2004], Lillesand et al. [2008], Ustin [2004]. This is commonly known as 'Earth observation from satellites'.

We should stress here that specific instruments are deployed for particular applications. For example, radar is mostly related to aerial traffic control and large scale meteorological data. Interferometric synthetic aperture radars, such as RADARSAT, TerraSAR-X, Magellan, are typically used to produce precise digital elevation models (DEM) of large scale terrain. On the other hand, laser and radar altimeters provide useful information of the sea floor, height and wave-length of ocean waves, wind speeds and direction, and surface ocean currents and directions. Another acquisition system such as the LIDAR (Light Detection And Ranging) is typically used to detect and measure the concentration of various chemicals in the atmosphere, while airborne LIDAR can measure heights of objects on the ground. A different family of technology is that of radiometers and photometers, which measure the reflected and emitted radiation in a wide range of wavelengths. They are typically used in meteorological applications and to detect specific chemical emissions in the atmosphere. One of the most widely used technology is that of thematic mappers, such as Landsat, whose popularity is due to the free availability of images and its continue and stable performance. They acquire images in several wavelengths of the electromagnetic spectrum (hence called *multispectral* mappers). They are mainly applied to updating the land cover and land use maps, and allows to identify particular materials, minerals, water or specific crops in the images.

1.1.4 THE REMOTE SENSING SYSTEM

All these kinds of physical processes, data, instruments, technologies and methodological frameworks define *The Remote Sensing System*, which can be organized in several parts, according to the taxonomies in [Campbell, 2007, Lillesand et al., 2008, Richards and Jia, 1999]:

1. *Energy Source*, which is devoted to the study of the Energy involved in the acquisition of the data.

2. *The Passive System*, which attends to the study of the solar irradiance from Earth's materials;

3. *The Active System*, which concentrates on the analysis of the irradiance from artificially generated energy sources, such as radar.

4. *Platforms* that carry the sensor, such as aircrafts, space shuttles, or satellites.

5. *Sensors*, detectors and devices to detect electromagnetic radiation (camera, scanner, etc.)

6. *Data Processing*, which involves techniques for enhancement, compression, data analysis, feature extraction, model inversion, image classification, etc.

7. *The Institutions*, which are in charge of the execution at all stages of remote-sensing technology.

In this chapter we will review the most important points of the Remote Sensing System. The *data processing* part is the core of the book, and dedicated chapters will deal with different parts of the image processing chain. Excellent books cover these topics in more detail too [Campbell, 2007, Liang, 2004, 2008, Lillesand et al., 2008, Mather, 2004, Rodgers, 2000]. Section 1.2 will review the foundations of the electromagnetic spectrum, and will give the definition of the main physical quantities involved in the image acquisition process. Section 1.2 is devoted to analyze the main properties and definitions of passive systems. Next, in Section 1.3, we revise the available platforms and sensors for optical remote sensing. Section 1.4 provides the reader with some of the most relevant institutions in the remote sensing community. We conclude in Section 1.5.

1.2 FUNDAMENTALS OF OPTICAL REMOTE SENSING

1.2.1 THE ELECTROMAGNETIC RADIATION

Electromagnetic radiation (EMR) travels through space in the form of periodic disturbances of electric and magnetic fields that simultaneously oscillate in planes mutually perpendicular to each other and to the direction of propagation through space at the speed of light ($c = 2.99792458 \times 10^8$ m/s). The electromagnetic spectrum is a continuum of all electromagnetic waves arranged according to frequency or wavelength, which are defined as the number of wave peaks passing a given point per second and the distance from peak to peak, respectively. Thus, both frequency, v (Hz), and wavelength, λ (m), of an EMR wave are related by its propagation speed, $c = \lambda v$.

The spectrum is divided into regions based on the wavelength ranging from short gamma rays, which have wavelengths of $10^{-6} \mu m$ or less, to long radio waves which have wavelengths of many kilometers. Since the range of electromagnetic wavelengths is so vast, the wavelengths are often shown graphically on a logarithmic scale (see Fig. 1.1 for a detailed classification of the electromagnetic spectrum). Visible light is composed of wavelengths ranging from 400 to 700 nm, i.e., from blue to red. This narrow portion of the spectrum is the entire range of the electromagnetic energy to which the human visual system is sensitive to. When viewed through a prism, this range of the spectrum produces a rainbow, that is, a spectral decomposition in fundamental harmonics or frequency components. Just beyond the red-end of the visible (VIS) region there is the region of infrared (IR) energy waves: near-infrared (NIR), shortwave-infrared (SWIR), middle-infrared (MIR), and the thermal-infrared (TIR).

The VIS and IR regions are commonly used in remote sensing. In particular, passive optical remote sensing is mainly focused in the VIS and NIR spectral region (VNIR), and in the SWIR since it depends on the Sun as the unique source of illumination. The predominant type of energy detection in the wavelength regions from 400 to 3000 nm (VNIR and SWIR) is based on the reflected sunlight.

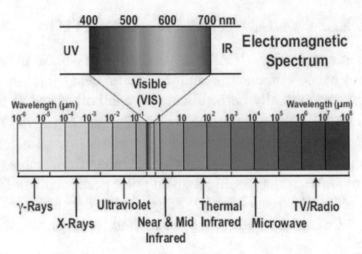

Figure 1.1: Electromagnetic spectrum classification based on wavelength range.

1.2.2 SOLAR IRRADIANCE

Energy generated by nuclear fusion in the Sun's core is the responsible for the electromagnetic radiation emitted by the Sun in its outer layer, which is known as the photosphere. It is the continuous absorption and emission of EMR by the elements in the photosphere that produces the light observed emanating from the Sun. The absorption characteristics of these elements produces variations in the continuous spectrum of solar radiation, resulting in the typical solar irradiance spectral curve. It must be stressed that 99% of the solar radiative output occurs within the wavelength interval 300-10000 nm.

The rate of energy transfer by EMR, the so-called *radiant flux*, incident per unit area is termed the *radiant flux density* or *irradiance* (W/m^2). A quantity often used in remote sensing is the irradiance per unit wavelength, and is termed the *spectral irradiance* (with units W/m^2/nm). The total radiant flux from the Sun is approximately 3.84 × 10^{26} W and, since the mean Earth-Sun distance is 1.496 × 10^{11} m, the total solar irradiance, over all wavelengths, incident at the Top of Atmosphere (TOA), at normal incidence to the Earth's surface, is

$$F_0 = \frac{3.84 \times 10^{26}}{4\pi(1.496 \times 10^{11})^2} = 1370 \text{ W/m}^2,\tag{1.1}$$

which is known as the *solar constant*, although it presents a considerable variation with time. The observed variations at the Sun are due to localized events on the photosphere known as *sunspots* and *faculae*[1]. An increased number of these events occurs approximately every 11 years, a period known as the solar cycle. However, the largest source of variation in the incident solar irradiance at the TOA is the orbit of the Earth around the Sun, due to the variable Earth-Sun distance that varies with the day of year.

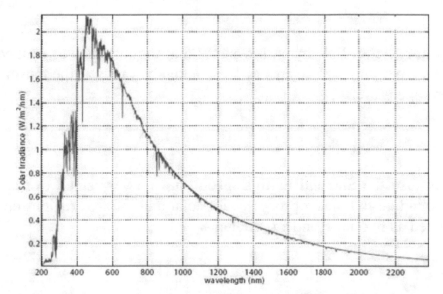

Figure 1.2: Solar spectral irradiance at the top of the Earth's atmosphere. See [Thuillier et al., 2003].

Space-borne instruments allow us measuring the spectral variation in solar irradiance at the TOA without the effects of the Earth's atmosphere which, depending on the wavelength of the radiation, can reduce the intensity of the measured radiation. Figure 1.2 shows the solar spectral irradiance at the top of the Earth's atmosphere [Thuillier et al., 2003].

It can be shown that the solar intensity curve resembles that of a Planck's distribution, $B(\lambda, T)$, for a blackbody at a temperature $T = 5777$ K (Fig. 1.3):

$$B(\lambda, T) = \frac{2hc^2}{\lambda^5 (\exp(\frac{hc}{k_B T \lambda}) - 1)}, \tag{1.2}$$

where h is the Planck's constant ($h = 6.626 \times 10^{-34}$ J s) and k_B is the Boltzmann's constant ($k_B = 1.38 \times 10^{-23}$ J/K). The maximum emission intensity of the curve occurs around 500 nm. This fact is consistent with Wien's displacement law, which states that the wavelength (λ_{max}) and the corresponding to the peak in Planck's curve for a blackbody radiating at a temperature T are related as follows:

$$\lambda_{max}T = 2.898 \times 10^6 \text{ (nm K)}. \tag{1.3}$$

Finally, the Stefan-Boltzmann law states that the total power emitted by a blackbody, per unit surface area of the blackbody, varies as the fourth power of the temperature:

$$F = \pi \int_0^\infty B(\lambda, T)d\lambda = 5.671 \times 10^{-8}T^4 \text{ (W/m}^2\text{)}. \tag{1.4}$$

Because the Sun and Earth's spectra have a very small overlap (Fig. 1.3), the radiative transfer processes for solar and infrared regions are often considered as two independent problems.

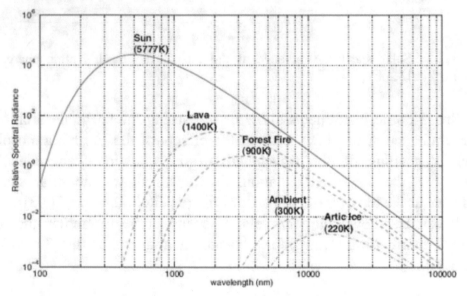

Figure 1.3: Blackbody emission of objects at typical temperatures.

1.2.3 EARTH ATMOSPHERE

The Earth's surface is covered by a layer of atmosphere consisting of a mixture of gases and other solid and liquid particles. The principal gaseous constituents, present in nearly constant concentration, are nitrogen (78%), oxygen (21%), argon (1%), and-minor constituents (<0.04%). Water vapor and an ozone layer are also present. The atmosphere also contains solid and liquid particles such as aerosols, water droplets (clouds or raindrops), and ice crystals (snowflakes). These particles may aggregate to form clouds and haze.

The vertical profile of the atmosphere is divided into four main layers: *troposphere*, *stratosphere*, *mesosphere*, and *thermosphere*. The tops of these layers are known as the *tropopause* (10 km), *stratopause* (50 km), *mesopause* (85 km), and *thermopause*, respectively. The gaseous materials extend to several hundred kilometers in altitude, though there is no well-defined limit of the atmosphere.

All the weather activities (water vapour, clouds, precipitation) are confined to the troposphere. A layer of aerosol particles normally exists near to the Earth's surface, and the aerosol concentration decreases nearly exponentially with height, with a characteristic height of about 2 km. In fact, the troposphere and the stratosphere together (first 30 km of the atmosphere) account for more than 99% of the total mass of the Earth's atmosphere. Finally, ozone exists mainly at the stratopause.

The characteristic difference between molecules and aerosols in the atmosphere is their respective size or radius [D'Almeida et al., 1991]. Molecules have a radius on the order of 0.1 nm, while aerosols can have a range of radii from 100 to 1000 nm. Both molecules and aerosols are optically active, causing the *absorption* and *scattering* of the electromagnetic radiation, respectively. Therefore, when the EMR from the Sun reaches Earth's atmosphere, it may be [Hapke, 1993]:

• *Absorbed*: incident radiation is taken in by the medium. A portion of the radiation is converted into internal heat energy that is emitted or radiated back at longer thermal infrared wavelengths.

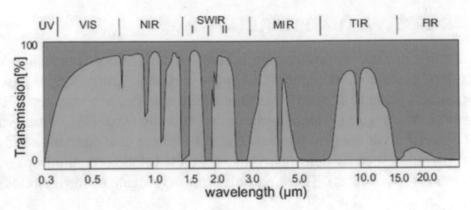

Figure 1.4: Relative atmospheric radiation transmission. Gray denotes absorption bands and blue areas denote atmospheric windows (transmission peaks).

- *Scattered*: incident radiation is dispersed or spread out by the particles suspended in the medium unpredictably in all directions. Radiation is absorbed and subsequently re-emitted at about the same wavelength without energy transformation, changing only the spatial distribution of the radiation.

- *Transmitted*: incident radiation passes through matter with measurable attenuation (absorbed or scattered).

- *Reflected*: incident radiation bounces off the surface of a substance in a predictable (specular reflection) or unpredictable (diffuse reflection) direction. Reflection consists in the scattering of the EMR by an object.

The overall effect is the removal of energy from the incident radiation. The amount of radiant energy that the atmosphere either removes or adds to that emitted or reflected from the Earth's surface depends on:

- the constituents of the atmosphere,

- the path length of radiation (function of the geometry of the illumination, the surface, and the observation),

- the reflectance of the surface target area and the surrounding scene.

Each type of molecule (constituent) has its own set of absorption bands in various parts of the electromagnetic spectrum [D'Almeida et al., 1991]. Absorption by atmospheric gases is dominated by that of water vapor (H_2O), carbon dioxide (CO_2), and ozone (O_3) with smaller contributions of the methane (CH_4), carbon monoxide (CO) and other trace gases. CO_2 and CH_4 are essentially uniformly distributed in the atmosphere, hence the effect of their absorption bands can be predicted reasonably well, while the water vapor distribution is rather variable in both location and altitude. Figure 1.4 shows the relative atmospheric radiation transmission of different wavelengths. A first consequence of the atmospheric effects is that wavelength bands used in remote sensing systems are usually designed to fall within these *atmospheric transmission windows*, outside the main absorption bands of the atmospheric gases, to minimize the atmospheric effects.

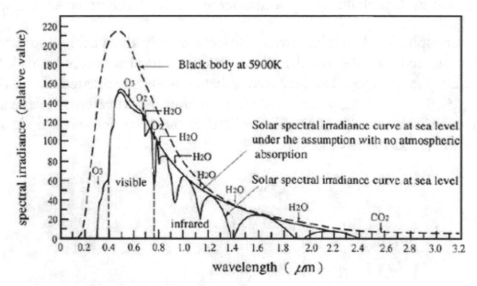

Figure 1.5: Solar irradiance at the top of the atmosphere and at the sea level, and blackbody emission spectrum at 5900 K.

Figure 1.5 shows the spectral features of the solar radiation outside the atmosphere (external line) and at the sea level (internal line). The maximum is located at $0.47\mu m$, being about 20% of the solar energy in wavelengths lower than that, and a 44% in the visible band, between $0.40\mu m$ and

$0.76 \mu m$. It is evident that water vapor is the most important absorber in the solar NIR spectrum, which contains about 50% of the solar energy.

1.2.4 AT-SENSOR RADIANCE

Signal measured at the satellite is the emergent radiation from the Earth surface-atmosphere system in the sensor observation direction. The incoming solar radiation, $F_0(\lambda)$, that we use for the observation of the surface, travels throughout the complex Earth atmosphere medium before is reflected by the surface, and the reflected signal travels again throughout the atmosphere before it arrives at the sensor. The measured at sensor radiance is called TOA radiance, and is the information we have to deal with when working with remote sensing data before atmospheric correction. The absorption and scattering processes affecting the solar electromagnetic radiation in its path across the atmosphere can be summarized as follows:

- Atmospheric absorption, which affects mainly the visible and infrared bands, reduces the solar radiance within the absorption bands of the atmospheric gases. The reflected radiance is also attenuated after passing through the atmosphere. This attenuation is wavelength dependent. Hence, atmospheric absorption will alter the apparent spectral signature of the target being observed.

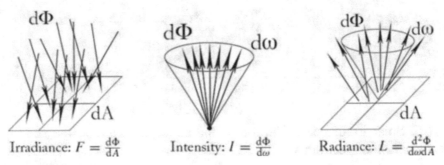

Irradiance: $F = \frac{d\Phi}{dA}$ Intensity: $I = \frac{d\Phi}{d\omega}$ Radiance: $L = \frac{d^2\Phi}{d\omega dA}$

Figure 1.6: Illustration of the geometric characterization of the incident irradiance, radiant intensity, and radiance.

- Atmospheric scattering is important only in the visible and near infrared regions. Scattering of radiation by the constituent gases and aerosols in the atmosphere causes degradation of the remotely sensed images. Most noticeably, the solar radiation scattered by the atmosphere towards the sensor without first reaching the ground produces a hazy appearance of the image. This effect is particularly severe in the blue end of the visible spectrum due to the stronger Rayleigh scattering for shorter wavelength radiation. Furthermore, the light from a target outside the field of view of the sensor may be scattered into the field of view of the sensor. This effect is known as the *adjacency effect*. Near the boundary between two regions of different brightness, the adjacency effect results in an increase in the apparent brightness of the darker region, while the apparent brightness of the brighter region is reduced.

We describe radiation in terms of energy, power, and the geometric characterization of power (Fig. 1.6). The radiant power, Φ, is the flux or flow of energy in the stream of time; hence, power is represented in watts (W). Flux density is the amount of radiant power emitted or received in a surface region. In fact, radiant *intensity*, *irradiance*, and *radiance* are different flux densities obtained by integrating the radiant power over the area, A, and/or the solid angle, ω, of the surface[2]:

- Irradiance, F, is defined as the received radiant power per unit area: $F = d\Phi/dA$ (W/m^2).

- Intensity, I, is an angular flux density defined as the power per unit solid angle: $I = d\Phi/d\omega$ (W/sr).

- Radiance, L, is an angular-area flux density defined as the power and unit area per unit solid angle (W/m^2/sr).

The fundamental radiometric quantity is the *radiance* that is the contribution of the electromagnetic power incident on a unit area dA by a cone of radiation subtended by a solid angle $d\omega$ at an angle θ to the surface normal. It has units W/m^2/sr, although *spectral radiance* (radiance per unit wavelength, W/m^2/nm/sr) is also commonly used, and is expressed mathematically as,

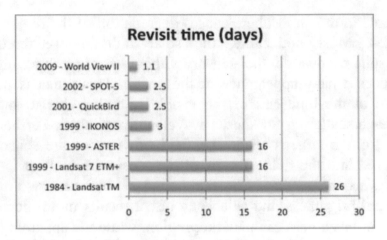

Figure 1.7: Revisiting time for representative satellite sensors.

$$L(\lambda, \theta, \psi) = \frac{\mathrm{d}^2 \Phi(\lambda)}{\cos(\theta) \mathrm{d}\omega \mathrm{d}A}. \tag{1.5}$$

where θ and ψ are the *zenith* and *azimuth* angle[3], respectively. Irradiance and intensity can be computed from the radiance, with appropriate integration: Irradiance is the integral of the radiance over all solid angles, and intensity is the integral of the radiance over all areas.

1.3 MULTI AND HYPERSPECTRAL SENSORS

This section reviews the main platforms and optical sensors for Earth Observation. Excellent reviews of other platforms and sensors, such as radar, thermal, LIDAR, or real aperture systems, can be found in [Campbell, 2007, Lillesand et al., 2008, Richards and Jia, 1999]. Passive optical remote sensing relies on solar radiation as the source of illumination. This solar radiation travels across the Earth atmosphere before being reflected by the surface and again before arriving at the sensor. Thus, the signal measured at the satellite is the emergent radiation from the Earth surface-atmosphere system in the sensor observation direction.

1.3.1 SPATIAL, SPECTRAL AND TEMPORAL RESOLUTIONS

In general, *resolution* is defined as the ability of an entire remote-sensing system, including lens antennae, display, exposure, processing, and other factors, to render a sharply defined image. Resolution of a remote-sensing image is of different types:

1. Spectral Resolution: of a remote sensing instrument (sensor) is determined by the bandwidths of the Electromagnetic radiation of the channels used. High spectral resolution, thus, is achieved by narrow bandwidths which, collectively, are likely to provide a more accurate spectral signature for discrete objects than broad bandwidth sensors.

2. Radiometric Resolution: is determined by the number of discrete levels into which signal radiance can be divided.

3. Spatial Resolution: in terms of the geometric properties of the imaging system, is usually described as the instantaneous field of view (IFOV). The IFOV is defined as the maximum angle of view in which a sensor can effectively detect electromagnetic energy.

4. Temporal Resolution: is related ot the repetitive coverage of the ground by the remote-sensing system. The temporal resolution of Landsat 4/5 is sixteen days. Figure 1.7 shows the revisiting time for some particular sensors.

1.3.2 OPTICAL SENSORS AND PLATFORMS

Examples of *multispectral* sensors on-board satellite platforms are Landsat/TM, SPOT/HRV, TERRA/ASTER or IKONOS, which present a few spectral bands and with broad bandwidths [Capolsini et al., 2003]. Recent satellite sensors are capable of acquiring images at many more wavelength bands. For example, the NASA's TERRA/MODIS [Salomonson et al., 1989] or ESA's ENVISAT/MERIS [Rast et al., 1999] sensors acquire tens of spectral bands with narrow bandwidths, enabling the finer spectral characteristics of the targets to be captured by the sensor. These kinds of sensors are commonly called *superspectral* sensors.

The earlier optical sensors considered a few number of bands, which readily demonstrated to be a limitation for detecting similar materials. A new class of imaging spectroscopy sensors, called hyper-spectral (imaging) sensors, acquire hundreds of contiguous narrow bands (or channels), and alleviate this problem. Hyperspectral sensors are a class of imaging spectroscopy sensors acquiring hundreds of contiguous narrow bands or channels. Hyperspectral sensors sample the reflective portion of the electromagnetic spectrum ranging from the visible region (0.4-0.7μm) through the near-infrared (about 2.4μm) in hundreds of N narrow contiguous bands about 10 nm wide[4]. Figure 1.8 shows the application of imaging spectroscopy to perform satellite remote sensing. In imaging spectroscopy or hyperspectral remote sensing [Goetz et al., 1985, Plaza et al., 2009, Schaepman et al., 2006], the resulting multispectral image consists of a simultaneous acquisition of spatially coregistered images, in several, spectrally contiguous bands, measured in calibrated radiance units, from a remotely operated platform.

The high spectral resolution characteristic of hyperspectral sensors preserves important aspects of the spectrum (e.g., shape of narrow absorption bands), and makes the differentiation of different materials on the ground possible. The spatially and spectrally sampled information can be described as a data cube (colloquially referred to as "the hypercube"), which includes two spatial coordinates and the spectral one (or wavelength). As a consequence, each image pixel is defined in a high dimensional space where each dimension corresponds to a given wavelength interval in the spectrum, $\mathbf{x}_i \in \mathbb{R}^N$, where N is the number of spectral channels or bands.

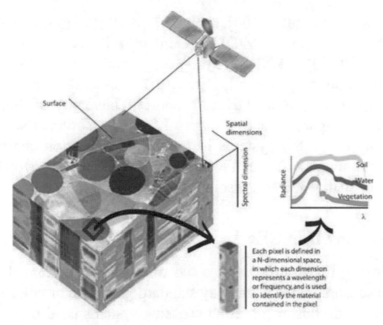

Figure 1.8: Principle of imaging spectroscopy.

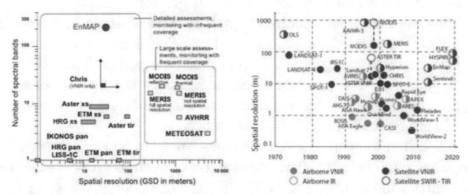

Figure 1.9: Left: Performance comparison of the main air- and space-borne multi- and hyperspectral systems in terms of spectral and spatial resolution. Credits: `http://www.enmap.de/`. **Right:** Evolution of the spatial-spectral resolution through the years.

There are experimental satellite-sensors that acquire hyperspectral imagery for scientific investigation such as ESA's PROBA/CHRIS [Barnsley et al., 2004, Cutter, 2004] and NASA's EO1/Hyperion [Ungar et al., 2003]. However, future planned Earth Observation missions (submitted for evaluation and approval) point to a new generation of hyperspectral sensors [Schaepman et al., 2006], such as EnMAP (Environmental Mapping and Analysis Program, GFZ/DLR, Germany) [Kaufmann et al., 2008, Stuffler et al., 2007], FLEX (ESA proposal) [Stoll et al., 2003], SpectraSat (Full Spectral Landsat proposal), HyspIRI (NASA GSFC proposal) [Green et al., 2008], ZASat (South African proposal, University of Stellenbosch), or HIS (Chinese Space Agency). Figure 1.9 shows a comparison of current multi- and hyperspectral systems in terms of information content, spatial resolution, and number of spectral bands. We also show the evolution of the technology in spatial-spectral resolution.

1.3.3 HOW DO IMAGES LOOK LIKE?

Remote sensing image processing need to create maps from the acquired data by the sensors, in the same way standard grayscale or color images are represented. To do this, most remote sensing systems need reference points on the ground to properly measure distances between spatially close acquired radiances. In a photographic image, for example, this kind of *correction* applied to the acquired signals is more accurate in the center of the image. The correction process in remote sensing data processing is known as *georeferencing*, and involves matching points, thus twisting and warping the image to produce accurate spatially meaningful images for analysis, processing and human interpretation, see Fig. 1.10. Currently, all data sold are typically provided georeferenced.

In addition to georefencing, the images need to be *radiometrically* and *atmospherically* corrected. The radiometric correction gives a scale to the pixel values (digital numbers, DN), e.g., the monochromatic scale of 0 to 255 will be converted to actual radiance values. The digitizing of data also make possible to manipulate the data by changing gray-scale values. The *atmospheric effects* are caused principally by the mechanisms of atmospheric scattering and absorption (cf. Section 1.2.3), and is one of the most challenging correction steps before image processing tools can be applied. As an example, Fig. 1.11 shows a MERIS FR Level-1 image before and after atmospheric correction.

1.4 REMOTE SENSING POINTERS

The widespread use of remote sensing technology is definitely due to a high number and activity of institutions aimed at developing, investing, studying, applying and teaching remote sensing. Particularly, one can find aeronautics organizations, research centers, universities, cartographic institutes, local/regional/national institutions, journal publishers, specialized conferences, companies, etc. In the following we briefly revise some of the most important ones. We are well aware of the incompleteness of the list, but we feel that this may be useful for a researcher coming from other fields.

Figure 1.10: From a raw image to the geocorrected AVIRIS image. The left figure shows an image constructed from a few bands in a hyperspectral cube acquired at low altitude. The data are rectangular, but the ground image is dramatically distorted due to aircraft motion as the sensor acquires cross tracks sequentially from top to bottom along the flight path. The middle image shows how the cross track actually registers to ground coordinates, The right image shows the corrected image: now the shapes on the ground are as expected, but the image borders are far from straight. Images courtesy of NASA/JPL-Caltech. Reproduced from Chapter 3 in [Camps-Valls and Bruzzone, 2009].

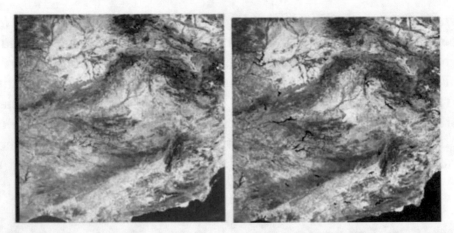

Figure 1.11: Top-Of-Atmosphere radiance from a MERIS FR half-swath product acquired on 14 July 2004 over Spain, and the surface reflectance map derived by the SCAPE-M atmospheric correction method [Guanter et al., 2008].

1.4.1 INSTITUTIONS

The most important *Aeronautics and Space* agencies are the European Space Agency (ESA), National Aeronautics and Space Administration (NASA), China National Space Administration (CNSA), Crew Space Transportation System (CSTS), Canadian Space Agency (CSA), Centre National d'Études Spatiales (CNES), British National Space Centre (BNSC), Agenzia Spaziale Italiana (ASI), Deutsches Zentrum für Luft- und Raumfahrt e. V. (DLR), and Instituto Nacional de Técnica Aeroespacial (INTA). Their web sites include information about ongoing missions, pointers to meetings and conventions, software and data resources.

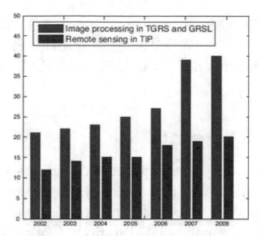

Figure 1.12: Number of papers in remote sensing journals (IEEE-TGRS and IEEE-GRSL) citing papers published in an image processing journal (IEEE-TIP). Source: Thomson Reuters' Web of knowledge, 2010.

1.4.2 JOURNALS AND CONFERENCES

Research is typically published in the following *Journals and Periodic Publications:* IEEE Transactions on Geoscience and Remote Sensing, IEEE Geoscience and Remote Sensing Letters, Remote Sensing of Environment, Photogrammetric Engineering & Remote Sensing, International Journal of Remote Sensing, Journal of Photogrammetry and Remote Sensing, and Applied Optics. Some applications and specific developments are encountered in IEEE Transactions on Image Processing, but the cross-fertilization between the remote sensing and the image processing communities has not become effective yet, see Fig. 1.12.

Concerning the most important conferences in the field, we should mention the IEEE International Geoscience & Remote Sensing Symposium, the SPIE Conference on Remote Sensing, International Society for Photogrammetry and Remote Sensing, Urban, Whispers, EARSeL Symposium & Workshops, and the EGU and AGU meetings.

1.4.3 REMOTE SENSING COMPANIES

Many companies are involved in the remote sensing data processing chain. Among them, the most well-known are GeoWeb Services, Analytical

Spectral Devices, ERDAS, Eurimage, Gamma AG, Remote Sensing and Consulting, GeoSearch, GEOSPACE, ImSAT, Brockmann Consult, Infoterra, Leica, Mathworks, OrbiSAT, SARMAP, SUN, ITT, ESRI, MapInfo, or ERMapper.

1.4.4 SOFTWARE PACKAGES

Remote sensing data is typically processed and analyzed with computer software packages. A number of popular packages are available: ERDAS, ESRI, ENVI, MapInfo, AutoDesk, TNTmips from MicroImages, PCI Geomatica by PCI Geomatics, IDRISI from Clark Labs, Image Analyst from Intergraph, RemoteView made by Overwatch Textron Systems, and eCognition from Definiens. In parallel, an increasing number of open source remote sensing software includes has appeared, among the most relevant ones are GRASS GIS, ILWIS, gvSIG, QGIS, OSSIM, Opticks, and the Orfeo toolbox. Very often, experiments are also conducted in IDL or Matlab by the scientific community. However, since some applications require managing large multidimensional images, the application-side of remote sensing image processing is conducted in specific software packages as the ones mentioned above.

1.4.5 DATA FORMATS AND REPOSITORIES

There exist several repositories to download remote sensing images, mainly linked to particular acquisition campaigns, international institutions (such as ESA or NASA), or companies. Some examples follow. Remote sensing image processing was mainly popularized by the freely available AVIRIS images, accessible from the NASA's website `http://aviris.jpl.nasa.gov/`. AVIRIS is an hyperspectral instrument that delivers calibrated images of the upwelling spectral radiance in 224 contiguous spectral channels (bands) with wavelengths from 400 to 2500 nanometers. Also NASA, at its Goddard Distributed Active Archive Centers (DAAC), provides data to perform studies on atmospheric dynamics, upper atmosphere, and global biosphere in `http://daac.gsfc.nasa.gov/`. The Earth Resources Observation

Systems (EROS) Data Center contains the world's largest collection of space and aircraft-acquired imagery of the Earth, available at `http://edcwww.cr.usgs.gov/`. Besides, some companies and institutions offer data to download, such as Astrium at `http://www.spotimage.com/`, Digital Global at `http://www.digitalglobe.com/`, the Canada Centre for Remote Sensing at `http://geogratis.cgdi.gc.ca/`, the USGS Global Visualization Viewer at `http://glovis.usgs.gov/`, or TerraServer at `http://www.terraserver.com/`. Recently, a few images in Matlab format are publicly available: this way researchers from other fields, such as computer vision, machine learning or image processing may have access to remote sensing imagery. For example, hyperspectral image examples are available at `https://engineering.purdue.edu/biehl/MultiSpec/hyperspectral.html`, `http://www.csr.utexas.edu/hyperspectral/codes.html` or `http://isp.uv.es`.

1.5 SUMMARY

The field of remote sensing is vast and embraces many fields of science, from physics to electronics, and from biology to computer science. In this introductory chapter, we have described the field of remote sensing and summarized its main aspects in terms of the key physical facts, sensors, platforms, and the communities involved. Even if incomplete, we believe that this review may help the beginner to find proper references for further information and to fairly follow the remainder of the book.

---

[1]*Sunspots* are dark areas on the photosphere which are cooler than surrounding regions. They have lifetimes ranging from a few days to weeks and are accompanied by strong magnetic fields. *Faculae* are regions of the photosphere which are hotter than their surroundings. They often occur in conjunction with sunspots and also possess strong magnetic fields and similar lifetimes.

[2]The *area* measures the surface region of a two- or three-dimensional object in square meters (m^2), while the *solid angle* is the projection of an area onto a sphere, or the

surface area of the projection divided by the square of the radius of the sphere, which is measured in stereoradians (sr).

[3]A zenith angle is a vector's angular deviation from an outward normal to the Earth's surface and azimuth is the horizontal angular variation of a vector from the direction of motion or true North.

[4]Other types of hyperspectral sensors exploit the emissive properties of objects by collecting data in the mid-wave and long-wave infrared (MWIR and LWIR) regions of the spectrum.

CHAPTER 2

The Statistics of Remote Sensing Images

This chapter studies the statistical characteristics of remote sensing images and compares them with those of standard photographic images. The issue of spatial, spectral and spatial-spectral redundancy is analyzed in detail through illustrative examples in real remote sensing images. The statistical behavior presented here is key in applications such as coding or restoration of remote sensing imagery.

2.1 INTRODUCTION

The historical origin of Earth Observation imagery is quite different from conventional photography. Artists have been using optical devices for image formation since long ago: in the 17th century Vermeer was using the *camera obscura* in his paintings [Gombrich, 1995] while his contemporary and neighbor Huygens was developing the wave theory of light [Hecht, 2001]. The advent of chemical photography and then digital photography may be considered just as improvements of the way visual information is stored (which had to be done by hand at Vermeer's *camera obscura*). The use of different (chemical or solid state) sensors tuned to different spectral bands in conventional photography was mainly motivated by the need to mimic human vision for non strictly scientific, but mainly artistic, purposes. The big deal for artists and film makers in the early 20th century was spatiotemporal resolution: color in movies came *later*. On the contrary, sensors in Earth Observation were developed for strictly scientific reasons. In the 19th and early 20th centuries, the spectrum of electromagnetic radiation coming from matter was the fundamental tool to obtain relevant physical information about the light sources, the reflecting surfaces and the media in between [Hecht, 2001]. According to this, it makes perfect sense that scientists dealing with problems on Earth Observation wanted engineers to develop sensors (and algorithms!) providing accurate spectral information. The big deal for scientists in those early days *was* the spectrum. First there was the *single pixel* spectrometer. Image (pixel-array) spectrometers came *later* [Goetz et al., 1985, Richards, 2005].

The above historical difference may roughly explain the different signal processing strategies used to deal with both kinds of data (conventional photography versus remote sensing imagery) in the last part of the 20th century. On the one hand, communities devoted to image processing and computer vision were mainly focused on spatiotemporal information, while color was sometimes seen as a convenient feature to add *afterwards*: for instance, not that long ago relevant image processing journals had to include tutorials on CIE colorimetry standards dating back to the 1930's [Sharma et al., 1998]. On the contrary, the remote sensing community was mainly focused on the spectral information while the spatiotemporal information was sometimes seen as a convenient feature to add *afterwards*: historically, the most widely used classification methods in hyperspectral imagery strongly rely in the intrinsically pixel-based *spectral signature* concept (e.g., see the taxonomies in [Camps-Valls et al., 2010b, Richards, 2005]).

Nonetheless, is there any fundamental (i.e. non historical) reason for these differences in the approaches? In other words, are these signals (scenes in conventional photography and remote sensing scenes) so statistically different? In this chapter we show that the basic answer is *no*: hyperspectral images coming from sophisticated satellite sensors are not that (statistically) different from conventional pictures that can be acquired with a cell phone. There are two fundamental reasons for this similarity: (1) the continuity and smoothness of the physical sources of the signals [Clarke, 1981, Singh et al., 2003] and (2) the self similarity of spatial structures across different spatial scales [Ruderman and Bialek, 1994]. Incidentally, it is worth remembering that the sources in conventional photography and Earth Observation imagery are actually the same, but viewed from different angles and distances. As will be explicitly shown below for hyperspectral remote sensing images, the spatio-spectral smoothness of the source also holds out of the visible range, so that statistical similarities between photographic scenes and remote sensing scenes also applies for the Ultra Violet and the Infra Red regions. In the same way, the self similarity of spatial structure across scales makes the statistical similarity to hold for very different spatial resolutions: from milimeters in conventional photography to kilometers in low spatial resolution remote sensing images.

That is good news for the image processing and computer vision communities, since the tools developed to process conventional images and video are expected to be successful (after minor tuning) when applied to the analysis of hyperspectral images: *the problem is not fundamentally different*. The above is even more striking because the increase of spatiotemporal resolution in Earth Observation sensors is leading to *vision-like* applications (e.g., monitoring, event detection). This may change the view from the physics-based *focus-on-the-spectrum* to the broader *focus-on-understanding* of the vision science and computer vision communities.

Of course, saying that *all* remote sensing scenes are *exactly the same* as conventional photography scenes is encouraging from the image processing perspective, but it is a bold overstatement. The physical image formation process can be qualitatively different (basically at wavelengths far from the visible spectrum, e.g., radar) thus changing the signal statistics (see, for instance, Fig. 2 in [Laparra et al., 2011a]). What we want to emphasize here is that the same signal processing techniques can be employed after the appropriate adaptation to the domain. Actually, as we will see in Chapters 3 and 4, many standard image processing techniques are adapted to the particular characteristics of remote sensing images.

In this chapter, we consider all the dimensions of the signal (the spectral dimension, the spatial dimension and the temporal dimension) as mathematically equivalent. This equivalence is conceptually convenient for a unified view of the problem; although with current computers, the joint consideration of all the dimensions at the same time implies a huge computational challenge, which leads us to the issue of properly characterizing the statistical regularities of the source in order to find more appropriate signal representations than the plain spectral-spatio-temporal domain. This issue, related to the more general concept of feature extraction, will be also visited in Chapter 3.

Even though noise statistics is highly relevant in a variety of applications, this chapter is intentionally focused on the scene statistics rather than in the noise statistics, which is an issue in its own right. For instance, beyond the conventional photography-like corruption [Aiazzi et al., 2002, Wettle et al., 2004], hyperspectral images can also present *non-periodic partially deterministic disturbance patterns* [Barducci and Pippi, 2001], which come from sensor specific image formation process and may require specific approaches to compensate them [Gómez-Chova et al., 2008]. Speckle [Hecht, 2001] is an issue in radar images [Solbo and Eltoft, 2004]. Artifacts of preprocessing, such as misregistration and compression, may have critical impact in change detection [Inglada et al., 2007] and classification [García-Vílchez et al., 2011] applications. However, providing a comprehensive list of sensor-dependent noise patterns is necessarily beyond the scope of an introductory book.

Despite we focus on the scene (and not the noise) statistics, note that really general restoration or image quality methods rely on the description of the statistical departure between the *noisy image* and a *typically behaved* image [Li and Wang, 2009, Simoncelli, 1997]. According to this, the statistical behavior of the hyperspectral scenes shown in this chapter can be used to discriminate between signal and noise as an extension of what it is proposed in [Elad et al., 2010, Li and Wang, 2009, Simoncelli, 1997].

This chapter is organized as follows. Section 2.2 analyzes second-order regularities in hyperspectral images and the shape of the marginal probability density function (PDF) in second order optimal representations, e.g., Principal Component Analysis (PCA). In Section 2.3, these spatio-spectral regularities are exploited in an efficient radiance coding example. This example stresses the convenience of jointly using the spatio-spectral information and its sparsity in local-frequency representations. Finally, Section 2.4 deals with higher-order regularities in hyperspectral images. To point out the generality of the considered statistical behavior, the examples in this chapter use real images of different spatial and spectral resolutions in different wavelength ranges: very high dimensional data acquired with the Infrared Atmospheric Sounding Interferometer (IASI) and the high spatial resolution Digital Airborne Imaging Spectrometer (DAIS) sensors [Chalon et al., 2001, Chang et al., 1993, Siméoni et al., 1997]).

2.2 SECOND-ORDER SPATIO-SPECTRAL REGULARITIES IN HYPERSPECTRAL IMAGES

The main point of this section is to stress the smoothness and the approximate Gaussianity of the different sources of the hyperspectral signal: the total radiance at the sensor, the sun irradiance, the transmittance of the atmosphere and the reflectance of the objects at the Earth's surface. The basic experimental results of the analysis performed here may be interpreted in terms of the classical result in [Clarke, 1981]: smooth sources with high correlation between samples analyzed using PCA give rise to DCT-like basis functions. In this section, we first study the spectral, the spatial and the spatio-spectral correlations, as well as the associated eigenfunctions and eigenvalues. Then, we consider the shape of the marginal PDFs in spatio-spectral PCA transformed representations, whose statistical characterization is relevant, for instance, in conventional denoising and coding applications [Gersho and Gray, 1992, Hyvärinen, 1999, Malo et al., 2000, Simoncelli, 1997]. Section 2.3 shows a coding application using the above second order statistical facts.

On the one hand, the fact that DCT-like basis functions are ubiquitous when using PCA in hyperspectral signals (no matter the considered dimension) tells us about the basic continuity of the physical sources. On the other hand, the good performance of the PCA-based transform coding example tells us about the fact that the Gaussian assumption (which is central in PCA-based transform coding [Gersho and Gray, 1992]) is approximately correct. However, as revealed by the shape of the marginal PDFs shown below, the Gaussian assumption (as expected) is not completely correct, which leads us to the analysis of higher order relations in hyperspectral signals.

2.2.1 SEPARATE SPECTRAL AND SPATIAL REDUNDANCY

Smoothness (second order redundancy) in the spectral and the spatial dimensions of hyperspectral images can be seen in the non-diagonal nature of the covariance matrices and in the width of the autocorrelation functions. To illustrate the spectral smoothness (second order redundancy) in remote sensing images, we use data come from the Infrared Atmospheric Sounding Interferometer (IASI) [Chalon et al., 2001, Siméoni et al., 1997], which provides 8461 spectral samples, between 3.62 and 15.5μm. Its spatial resolution is 25 km at nadir with an instantaneous field of view size of 12 km at a satellite altitude of 819 km. This data will be used again in Chapter 6. Figures 2.1 and 2.2 illustrate the spectral smoothness (for a fixed spatial position) of (1) the irradiance at the IASI sensor, (2) the solar irradiance at the Earth's surface and (3) the atmosphere transmittance. Note that this spectral smoothness holds at different wavelength ranges: visible and near infrared in the Sun and atmosphere examples, and far infrared in the IASI data.

The spectral irradiance of the sunlight at the Earth's surface varies from reddish (sunrise and sunset) to blueish (midday) due to the Rayleigh effect. This dependence has been usually approximated as a Planckian radiator (see Chapter 1) with correlated color temperatures in the range [5500, 6500] K, see [Stiles and Wyszecki, 1982]. At the same time, the total integrated irradiance reaches its peak at midday. In order to simulate such effects for this illustration, we randomly selected noisy Planckian-like spectra corresponding to correlated color temperatures in the above range [Krystek, 1985]. Moreover, the generated spectra are scaled such as the integrated irradiance is related to the color temperature. Figure 2.2a shows samples generated in this way along with the corresponding colors and the Planckian locus in the CIE xy diagram (Fig. 2.2b). This diagram is useful to show the color variation of the spectrum samples Stiles and Wyszecki [1982] Note that the average of the samples (in red) is similar to the CIE D65 illuminant which is the usual model for daylight [Stiles and Wyszecki, 1982]. Figure 2.2c shows the first six spectral eigenvectors of such ensemble, with the usual result for smooth functions: oscillating functions of increasing frequency for decreasing eigenvalues. Figure 2.2d shows the spectral eigenvectors from atmosphere aerosols reported in [Timofeyev et al., 2003]. Again, the result consists of oscillating functions of increasing frequency for decreasing eigenvalues.

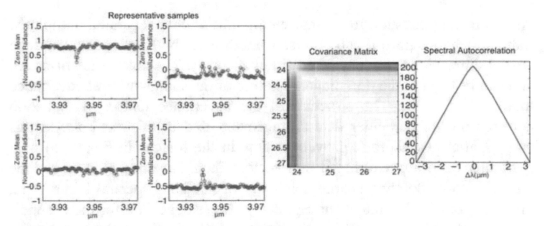

Figure 2.1: Irradiance at the sensor is spectrally smooth: IASI spectra samples, spectral covariance matrix and spectral autocorrelation. Zero mean normalized radiance is given by $\hat{L}_\lambda = \frac{L_\lambda - \bar{L}_\lambda}{\sigma(L_\lambda)}$, where L_λ is the absolute spectral radiance, $\bar{L}_\lambda$ is the average spectral radiance, and $\sigma(L_\lambda)$ is its the standard deviation. Note that the global mean was removed from the whole hyperspectral cube so that individual samples such as those in the examples do not have zero mean.

Figure 2.3 shows spatial samples (for a fixed spectral band, $\lambda = 4.42\mu m$) of radiances measured using IASI, the corresponding covariance matrix and the (wide) autocorrelation function. Similar behavior (wide autocorrelation) is obtained using a very different spatial resolution in the visible wavelength region (DAIS data [Chang et al., 1993]), see Fig. 2.5.

Summarizing, we observed that both spectral and spatial remote sensing image features are essentially smooth (e.g., they show a strong second order redundancy), show wide autocorrelation functions or extremely non-diagonal covariance matrices. These *separate* results on the spectral and spatial characteristics justify the *separate* use of DCT and wavelet-based transforms in the spectral and the spatial dimensions for hyperspectral radiance coding as is usually done in the current literature [Penna et al., 2007]. Besides, they are consistent with the corresponding previously reported results for conventional photographic scenes [Hancock et al., 1992, Maloney, 1986].

2.2.2 JOINT SPATIO-SPECTRAL SMOOTHNESS

An interesting question to be answered is if, not only separated but jointly, the whole hyperspectral cube is also smooth when both the spectral and the spatial dimensions are considered simultaneously. The answer is afirmative and can be seen in the non-diagonal nature of the spatio-spectral covariance matrix (Fig. 2.4b). The structure of the covariance matrix in Fig. 2.4b depends on the particular data arrangement used in this example. In this case, spatial neighbors (e.g., yellow dots in the images in Fig. 2.4a) of a particular band are rearranged in a vector. Then, the same process is done with the corresponding spatial neighbors of neighbor spectral bands (e.g., green regions in the images in Fig. 2.4a); this is done first for the channels in one of the broad bands (e.g., the images in the purple bracket), and so on (for the images in the orange and pink brackets). As a result, different blocks in the covariance matrix indicate different relations: spatial relations within a single spectral channel (e.g., yellow squares in the covariance), relations between values in neighbor spatial locations and neighbor spectral channels (e.g., green squares in the covariance), and relations between broad bands (purple, orange and pink squares in the covariance). The structure in the covariance shows that second order relations decrease with spatio-spectral distance, and these relations are more intense within broad bands. According to this, and in order to save computation time (but without loss of generality), in the following illustrations (here and in Section 2.3) we will restrict ourselves to the first broadband region (top left green squares submatrix of the covariance).

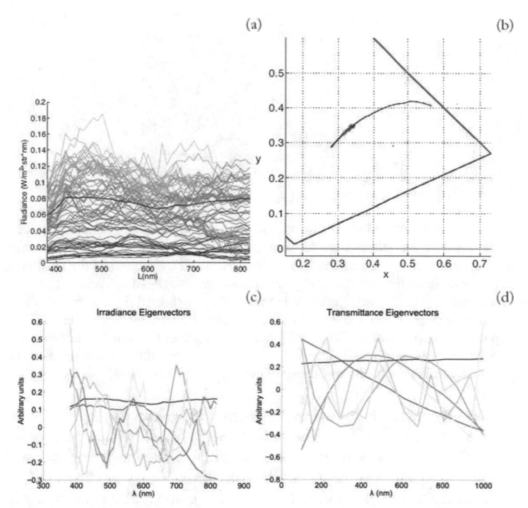

Figure 2.2: Sun irradiance and atmosphere transmittance (aerosol contribution) are spectrally smooth: (a) Irradiance samples (in different gray level according to their total energy). In red is shown the average, which is similar to the standard CIE D65 spectrum that accounts for usual daylight. (b) Irradiance samples in the CIE xy color space Stiles and Wyszecki [1982] in different gray levels according to their luminance. In red the Planckian locus. (c) First eigenvectors of the Sun irradiance. (d) First eigenvectors of the atmosphere transmittance. In (c) and (d), darker line style means higher eigenvalue.

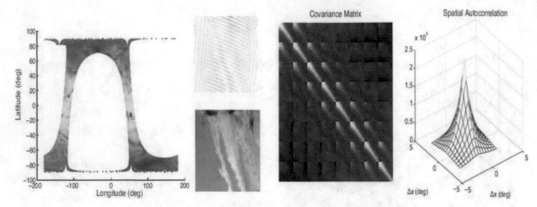

Figure 2.3: Hyperspectral scenes are spatially smooth: IASI spatial samples (complete satellite orbit at the left and spatial samples and interpolated image at the center), spatial covariance matrix and spatial autocorrelation. The structure of the covariance matrix comes from the fact that 2D spatial blocks are rearranged as vectors. Note that, as in the spectral case, autocorrelation is very wide as expected from a smooth signal. It is important to stress the fact that smoothness (intuitively visible in the non-interpolated image showing Baja California) is independent from the interpolation: interpolation was performed just for mathematical convenience but it does not substantially affect the qualitative decay of correlation with physical distance.

The eigenfunctions of the spatio-spectral covariance for the data in this region is shown in Fig. 2.4c. The smoothness revealed by the non-diagonal nature of the covariance matrix gives rise to DCT-like basis functions with increasing frequency as the eigenvalues decrease. This was expected from the classical DCT-PCA result in [Clarke, 1981], and it also consistent with the equivalent results for conventional photography in Doi et al. [2003] and [Singh et al., 2003].

We would like to stress that this behavior is not specific of IASI data (very low spatial resolution in the far infrared), but it can also be observed for other sensors with very different spatial and spectral resolutions. As an illustration, Figure 2.5 shows sample images acquired by the DAIS sensor in different bands of the visible range and the same bands for the corresponding spatio-spectral basis functions arranged as in the example of Fig. 2.4c.

Images in this example were collected by the Digital Airborne Imaging Spectrometer (DAIS 7915) in the HySens project [Dell'Acqua et al., 2004]. The spatial resolution is of 2.6 meters per pixel, and the scene size covers nearly one square km. DAIS covers 79 spectral channels in the visible and infrared wavelengths [Chang et al., 1993]; however, in this illustration, we limit ourselves to the visible range just for computational convenience.

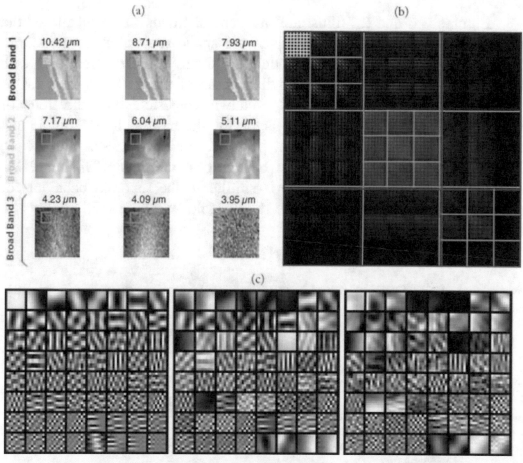

Figure 2.4: Hyperspectral scenes are spatio-spectrally smooth (IASI data). Top left: IASI spatial samples for different bands. Top right: joint spatio-spectral covariance matrix. Bottom row: spatio-spectral DCT-like basis functions of the eigendecomposition of the first broadband. In this bottom row, the panels show the spatial pattern (8 × 8 pixels) of each basis function for the 10.42μm band (left), the 8.71μm band (center), and the 7.93μm band (right). In these plots, the functions are sorted from left to right and top down as a function of the eigenvalue.

Again, this *joint* spatio-spectral smoothness justifies the use of *joint* 3D local frequency transforms in the spectral and the spatial dimensions for hyperspectral radiance coding, as is usually done in the current literature [Penna et al., 2007]. Moreover, this is consistent with the results previously reported for conventional photographic scenes [Doi et al., 2003, Wachtler et al., 2001].

Figures 2.6 and 2.7 illustrate the general trends of the shape of the marginal PDFs for the coefficients of the spatio-spectral PCA transform. Marginal PDFs are highly kurtotic and their variance decreases with frequency: see the progression of the widths of the PDFs in Fig. 2.6 as higher coefficients are considered, which is consistent with the trend of variance in Fig. 2.7a. The shape factor for the marginal PDFs is almost constant for every PCA coefficient [Gersho and Gray, 1992, Malo et al., 2000], as seen in Fig. 2.7b. This behavior of marginal PDFs in PCA-decorrelated hyperspectral signals is similar to the reported behavior for photographic images [Hyvärinen, 1999, Malo et al., 2000, Simoncelli, 1997].

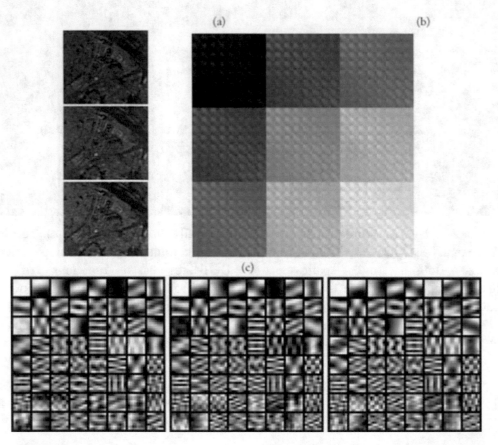

Figure 2.5: Hyperspectral scenes are spatio-spectrally smooth (DAIS data). (a) Sample images at fixed spectral bands in the visible range for the DAIS sensor: 500, 600 and 700 nm from top to bottom. (b) Submatrix of the spatio-spectral covariance matrix for the considered bands. (c) Spatial pattern (8 × 8 pixels) of each basis function for the 500nm band (left), the 600nm band (center), and the 700nm band (right). In these plots, the functions are sorted from left to right and top down as a function of the eigenvalue. Note that just a submatrix of the covariance is shown for better visualization, while the eigendecomposition is performed on the whole covariance.

Note also that, as reported in photographic images [Hyvärinen, 1999, Olshausen and Field, 1996, Simoncelli, 1997], the non-Gaussian nature of the marginal PDFs implies that the joint PDF is also non-Gaussian. According to this, even though the Gaussian assumption may be appropriate enough in some applications (see the coding example in the next section), it is necessary to consider higher order statistics to fully characterize the behavior of hyperspectral data.

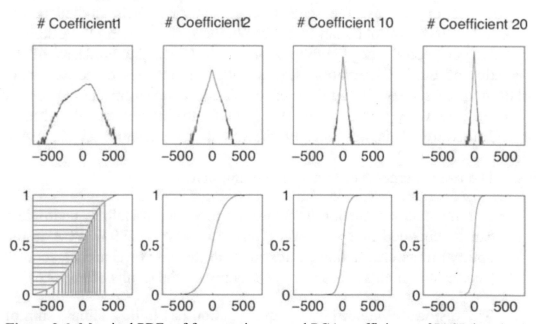

Figure 2.6: Marginal PDFs of four spatio-spectral PCA coefficients of IASI data (top), and the corresponding sigmoid functions for optimal quantizer design (bottom). Probability axes are in log-scale. See comments on quantization in Section 2.3.

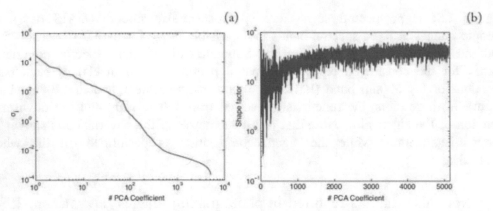

Figure 2.7: Standard deviation of spatio-spectral PCA coefficients (a), and shape factor of the marginal PDFs (b), of IASI data. See [Gersho and Gray, 1992, Malo et al., 2000] for the definition of the shape factor and its role in optimal bit allocation, details in Section 2.3.

2.3 APPLICATION EXAMPLE TO CODING IASI DATA

In this section, we use the second order statistical facts studied before together with the kurtotic marginal PDFs in the spatio-spectral PCA domain in an hyperspectral image coding application. Given the similarity of the statistics of Earth Observation scenes with the statistics of conventional photographic scenes, the application of classical transform coding results [Gersho and Gray, 1992, Wallace, 1991] to the hyperspectral cubes is straightforward. This was already showed in [Abousleman et al., 1995, Saghri et al., 1995].

The basic elements of transform coding involve:

- A redundancy reduction transform so that each transform coefficient can be encoded independently [Gersho and Gray, 1992]. PCA is the optimal transform in that context only in the case of Gaussian sources (see the comments about non-Gaussianity on the results of Fig. 2.6).

- The distortion induced by the quantization has to be a simple sum of the distortions in each coefficient [Gersho and Gray, 1992]. This is only possible if the meaningful distortion metric is diagonal in the transform domain as discussed in [Camps-Valls et al., 2008b, Epifanio et al., 2003, Malo et al., 2006]. Mean Square Error (MSE) together with orthogonal transforms (such as PCA) fulfill this requirement. However, note that MSE is not always a meaningful distortion metric in remote sensing applications [Alparone et al., 2004].

• Optimal quantization of each coefficient has to take into account its marginal PDF [Gersho and Gray, 1992, Lloyd, 1982]. Bottom row in Fig. 2.6 illustrates this issue: the quantization coarseness (orange lines in the example) has to be finer in the regions with bigger population (around zero in these kurtotic PDFs). This amplitude dependent quantization is obtained from the marginal cumulative density functions. Moreover, if the meaningful metric is amplitude dependent, these basic results are modified [Malo et al., 2000].

• Optimal bit allocation among transform coefficients has to take into account the variance of the coefficients and the shape factor of the marginal PDFs [Gersho and Gray, 1992]. Figure 2.6 illustrates the elements that shape bit allocation in PCA transform coding of hyperspectral data according to an MSE criterion. More meaningful distortion metrics in the transform domain (even the simpler, diagonal ones) modify the bit allocation result by weighting the variance and the shape factor [Malo et al., 2000].

According to this, the results in the previous section are needed in optimal coding schemes.

The coding scheme of IASI radiances currently used in EUMETSAT data processing schemes is based on the truncation of the spectral PCA [Hultberg, 2009] (block diagram in pink in Fig. 2.8). In the following example, different improvements are considered either related to (1) the consideration of the spatial information, or (2) the consideration of the marginal PDFs and coefficient variance (eigenvalues) for optimal bit allocation, according to the optimal transform coding elements listed above (see the rest of block diagrams in Fig. 2.8).

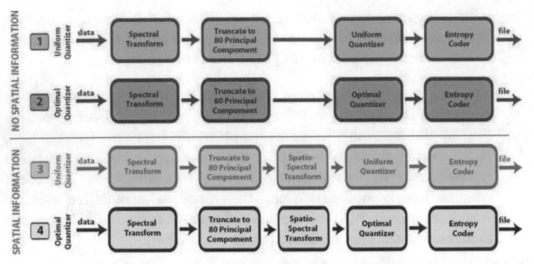

Figure 2.8: Compared schemes in coding IASI radiances. The colors of these block diagrams (pink, red, blue and black) will be used in the following figures to identify the results of each schemes.

Top panel in Fig. 2.9 shows the rate-distortion performance of the different approaches compared in Fig. 2.8. An appropriate bit allocation in the transform domain (red and black) is better than a simple truncation and uniform quantization (pink and blue). More important (in MSE terms) is the exploitation of spatial regularities: note the reduction of asymptotic distortion when using this information (blue and black curves) with regard to the spectral-only case (pink and red curves). The bottom panel in Fig. 2.9 shows representative results of the reconstructed spectra for a fixed spatial position and the four considered approaches. Different colors indicate the different approaches and different line style indicate different compression rate. Results corresponding to 0.15, 0.45 and 0.70 bits/datum are shown and compared to the true spectrum (in green). As expected, increasing the bit rate leads to better reconstructions, but note that some approaches converge to the actual signal faster (at lower bit rates) than others, consistently to the results in the top panel. Figure 2.10 shows representative results of the reconstructed spatial patterns for a fixed spectral band and the four considered approaches. Results corresponding to the range [0.01, 1.11] bits/datum are shown and compared to the true image (in the top left). Note that the better the coding scheme (consideration of spatial information and appropriate quantizer), the lower bitrate needed for comparable distortion, consistently to the rate-distortion results.

These results show that (1) the consideration of the spatial regularities is the key to achieve better coding results, and (2) the appropriate consideration of the marginal PDFs for quantizer design allows us to obtain the asymptotic distortion (imposed by the dimensionality reduction) for a smaller number of bits (see the decays of distortion in Fig. 2.9).

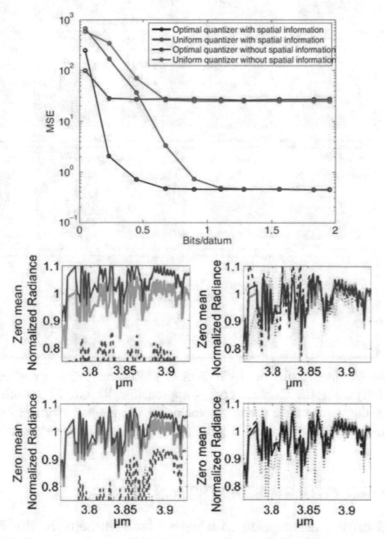

Figure 2.9: Coding results on IASI data (I). Top panel: rate distortion. Line colors refer to the different schemes identified by the same colors in Fig. 2.8. Bottom panel: reconstruction of representative spectra (ground truth in green) using the different

approaches (using the same color code as in previous figures). Line style represents the reconstruction using a particular bit rate: dotted, dashed and solid lines refer to 0.15, 0.45 and 0.70 bits/datum, respectively.

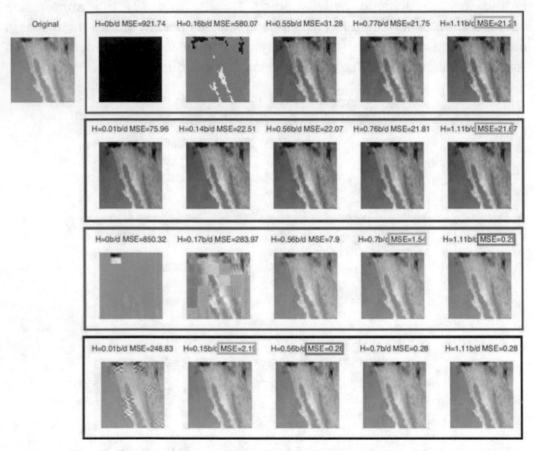

Figure 2.10: Coding results on IASI data (II). Reconstruction of a representative image (top left) using the different coding approaches. Results from each scheme are identified by the color of the rectangle enclosing the images. Numbers show the bit rate and MSE in each solution. Highlighted numbers show comparable distortion values.

2.4 HIGHER ORDER STATISTICS

PCA relies on the assumption of a particular symmetry of the PDF of the underlying phenomenon [Laparra et al., 2011b]. As stated above, PCA-based transform coding assumes Gaussian PDFs [Gersho and Gray, 1992].

However, kurtotic results of the marginal PDFs in the spatio-spectral PCA domain (Fig. 2.6) clearly show that Earth Observation scenes are non-Gaussian. This is bad and good news at the same time. The bad news is that second order characterization is not enough to fully describe a non-Gaussian source. Techniques exploiting or describing higher order relations will be needed, such as Independent Component Analysis (ICA) [Comon, 1994, Hyvärinen et al., 2001]. The good news is that the sparsity found in the coefficients of remote sensing scenes when represented in local frequency transforms is an additional feature of the scenes that can be used as a proper characterization of the signal. This opens the possibility of imposing this sparsity as a prior in applications where missing information has to be recovered [Bobin et al., 2009], as done when dealing with conventional photographic images [Elad et al., 2010]. An active field in remote sensing deals with the sparse representation of the images, and will be reviewed in Chapter 5.

The similarity between the remote sensing images and the conventional photographic images has also been shown to hold for the linear ICA results [Birjandi and Datcu, 2010]: the independent features will be edge filters similar to wavelets, as previously reported in [Bell and Sejnowski, 1997, Olshausen and Field, 1996]. The main physical reason to the equivalence is that in hyperspectral scenes spatially stationary regions are separated also by edges. This equivalence is confirmed by the similar conditional probability results in spatio-spectral wavelet decompositions of hyperspectral data that will be shown later.

However, the general question is: do linear redundancy reduction transforms provide a full description of the PDF at hand? In other words, can linear techniques, such as PCA and ICA (or their approximations, local DCT and wavelets), remove all the statistical relations between the transform coefficients? Here we show that (as in conventional photography) additional, necessarily higher-order, relations do exist between the coefficients of these scenes in spatio-spectrally decorrelated domains.

Figure 2.11 shows the existence of these relations in two different ways: (1) pointing out that the conditional PDF of neighbor coefficients reveals inter-coefficient redundancy, and (2) measuring mutual information among spatio-spectral wavelet coefficients. The horizontally oriented bow-tie shape of these conditional probabilities means that the knowledge of the value of one coefficient does not tell anything about the *mean* of the neighbor coefficients (note that the average of coefficient w_j is independent of the value of coefficient w_i). The linear wavelet transform (assumed to be similar to linear ICA) does remove this dependence. However, the increase of the variance of w_j as a function of w_i reveals that there is a relation between the *energies* of the coefficients. This reveals a failure of the basic linear ICA model: hyperspectral signals do not strictly come from a linear combination of independent sources. Even though linear ICA takes into account relations of order higher than two, this is not enough to fully describe the signals either.

These results (relations and dependence with distance) are consistent with equivalent results in conventional photography on the the correlation between the energy of the PCA (local DCT, local Fourier and wavelet) coefficients [Gutiérrez et al., 2006, Malo et al., 2006], the classical bow-tie shape of the conditional PDFs [Hyvarinen et al., 2003, Simoncelli, 1997], and the mutual information results in 2D wavelets [Buccigrossi and Simoncelli, 1999, Laparra et al., 2010, Liu and Moulin, 2001, Malo and Laparra, 2011].

These higher-order properties of hyperspectral images justify the exploitation of relations among coefficients in different subbands in spatio-spectral wavelet decompositions as done in recent coding applications [Dragotti et al., 2000, Penna et al., 2007, Tang and Pearlman, 2005]. The non-Gaussian nature of the remote sensing signals revealed by the highly kurtotic structure of the marginal PDFs after local frequency transforms (as in Fig. 2.6) is also the ultimate justification for the use of wavelet shrinkage techniques in denoising applications. Again, the success of these denoising techniques in hyperspectral images [Chen and Qian, 2011, Letexier and Bourennane, 2008] is actually related to the use of the same properties of photographic scenes [Donoho, 1995, Hyvärinen, 1999]. The higher-order similarities between Earth observation images and photographic scenes seen in Fig. 2.11 suggest that recent advances in image coding and denoising in photographic images based on the same facts [Camps-Valls et al., 2008b, Gutiérrez et al., 2006, Laparra et al., 2010, Malo et al., 2006] could be successfully extended to hyperspectral imagery.

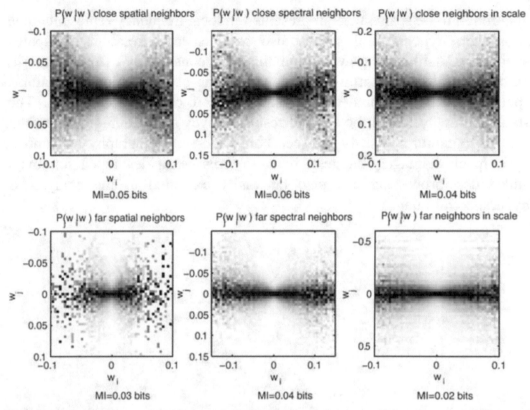

Figure 2.11: Higher-order relations in hyperspectral data: conditional probabilities and Mutual Information (MI) between neighbor coefficients of spatio-spectral wavelet transforms of IASI data. The bow-tie shape indicates that, while the average of the second coefficient w_j (vertical axis) is independent from the average of the first coefficient w_i (horizontal axis), as expected from decorrelation, its variance (energy) is not independent. The considered neighbors are in space (left), wavelength (center) and scale (right). Top row shows the conditional PDFs for close neighbors (strong dependence) while the bottom row shows the same thing for distant neighbors (weaker dependence), as consistently indicated by the MI values.

2.5 SUMMARY

In this chapter we analyzed the second-order and the higher-order statistical regularities of hyper-spectral remote sensing images to point out that they are similar to those found in conventional photography. Similarly, to natural images, hyperspectral scenes are smooth: the values in neighbor locations and neighbor wavelengths are highly correlated. Hyperspectral scenes are also found to be sparse when represented in local frequency representations. Moreover, hyperspectral scenes also show specific relations in spatio-spectral, ICA-like, 3D wavelet domains. The consistency of this similarity across spatial and spectral resolutions and the usefulness of the (natural) spatio-spectral approach was illustrated in a coding application. This statistical analysis and the applied results convey an encouraging message: the fundamental similarity between conventional photography and optical hyperspectral imagery suggests that the tools being developed for image and video processing can also be easily extended to the analysis of hyperspectral images.

CHAPTER 3

Remote Sensing Feature Selection and Extraction

This chapter reviews the important fields of feature selection and extraction of remote sensing images. These steps are of paramount importance in many applications: from the posterior image segmentation and object location, to properly estimate physical variables from the images. A wide range of standard techniques have been adopted by the remote sensing community but, at the same time, many of them have been necessarily adapted to the particular features of images. In addition, the issue of extracting physically-meaningful features from images is an active field. We will review the main contributions in the area through an extensive literature review and real examples.

3.1 INTRODUCTION

A major problem in remote sensing image processing is the huge amount of data involved. For instance, NASA's Airborne Visible Infra-Red Imaging Spectrometer (AVIRIS) is able to cover the wavelength region from 0.4 to $2.5 \mu m$ using 224 spectral channels, at a nominal spectral resolution of 10 nm. Hyperspectral sounders such as IASI used in Chapter 2 acquires more than 8000 spectral channels. Statistical redundancy is definitely an issue, as we revised in the previous chapter. From a pure information-theoretic approach, the relevance of the spectral channels acquired by remote sensing instruments can be analyzed. But probably, statistical approaches may miss something when selecting features in terms of physical meaning. Selecting the most relevant features, either spectral or a specific number of spatial-spectral components, is of paramount importance for a successful application of posterior classification and regression processing steps.

In this scenario, the curse of dimensionality [Bellman, 1961], also known as Hughes' phenomenon [Hughes, 1968], refers to the problems associated with multivariate data analysis as the dimensionality increases. This problem is specially relevant in remote sensing since, as new technologies improve, the number of spectral bands and the spatial resolution are continuously increasing. There are two main implications of the curse of dimensionality that critically affect image processing applications in remote sensing: there is an exponential growth in the number of examples required to maintain a given sampling density (e.g., for a density of n examples *per* bin with d dimensions, the total number of examples should be n^d); and there is an exponential growth in the complexity of the target function (e.g., a density estimate or a classifier) with increasing dimensionality. The close relationship between the complexity of the models and the number of available samples suggests the idea of a prior reduction of the input space dimensionality in order to avoid wrong estimations of model parameters [Fukunaga and Hayes, 1989]. This objective can be achieved in two different ways. The first one is to identify those variables that do not contribute to the predefined task and to ignore them (*feature selection*). The second approach is to create a subset of new features by combinations of the existing features, i.e., to find a transformation to a lower dimensional feature space while retaining the useful information for the problem (*feature extraction*). The following sections describe the most widely used feature selection and feature extraction methods in the remote sensing literature.

3.2 FEATURE SELECTION

Feature selection is usually performed using generic algorithms that are applied in a huge variety of fields. These algorithms can be applied directly to the observations, in both the spectral and the spatial domains, or to further extracted features, as will be described in Section 3.3. However, in the field of remote sensing, feature selection methods have several advantages compared to feature extraction when applied to the selection of spectral features: 1) *data transmission*, since only selected channels might be transmitted from satellite; 2) *interpretability of results*, since the selected features are spectral bands with physical meaning; and 3) *extrapolation of results* to other instruments with slightly different spectral band configurations. The main risk is the loss of information if the feature selection is wrong.

Reducing the dimensionality of the data while keeping the most of its expressive power is the goal of feature selection, and a great number of methods have been proposed in the remote sensing literature. The feature selection problem (FSP) in a 'learning from samples' approach can be defined as choosing a subset of features that achieves the lowest error according to a certain loss functional. Following a general taxonomy, the FSP can be tackled using *filter* [Blum and Langley, 1998] and *wrapper* [Kohavi and John, 1997] methods:

Filter methods use an indirect measure of the quality of the selected features, e.g., evaluating the correlation or dependence between each input feature and the observed output.

Wrapper methods directly optimize a fitness criterion between the inputs and the outputs provided by the learning machine under consideration, e.g., a neural network. This approach guarantees that, in each step of the algorithm, the selected subset improves performance of the previous one in terms of accuracy.

It is worth noting that filter methods might fail to select the right subset of features if the used criterion deviates from the one used for training the learning machine, whereas wrapper methods can be computationally intensive since the learning machine has to be retrained for each new set of features.

3.2.1 FILTER METHODS

Filters converge much faster than wrapper methods, and select features independently of the subsequent classifier, which facilitates feature interpretation. Note that often the remote sensing user is much more interested in *understanding* the relative relevance of the considered features than in minimizing a classification error.

Filter methods have been extensively studied in remote sensing. In [Serpico and Moser, 2007], a feature selection procedure was proposed to combine spectral channels, while a canonical correlation method was applied in [Paskaleva et al., 2008] to sensors with overlapping bands. Strategies to constrain the search space were deployed in [Serpico and Bruzzone, 2001]. In [Camps-Valls et al., 2010a], a filter approach based on measuring the nonlinear dependence between the features and the class labels was introduced. The proposed method was successfully compared to several standard and state-of-the art methods. Lately, also graph-based semisupervised feature selection approaches have been proposed for classification of remote sensing images. In [Chen et al., 2010b], the authors evaluate the class separability of unbalanced sample sets while including local information from a graph to assess the ability of selected features in preserving the geometrical and discriminant structures.

These feature ranking and feature selection algorithms are independent of the learning machine subsequently used to obtain results. They filter out features that are not useful attending to a performance evaluation metric calculated directly from the data, without direct reference to the results of any learning machine. For example, a standard dimensionality reduction strategy consists in eliminating redundant information, by means of local correlation criterion between contiguous spectral bands, and then performing a selection of the most discriminative features based on combination of several feature ranking methods.

3.2.2 WRAPPER METHODS

The main problems with most of the filter methods are that: 1) the relationship between sets of features and the class labels is not jointly considered, as they are usually pairwise feature-label measurements, and 2) they do assume either linear dependencies (e.g., by using Pearson's correlation) or *ad hoc* criteria of class separability (e.g., mean class differences). These problems reduce their usefulness and have motivated the recent interest in feature selection methods wrapped around a predictor, whose accuracy is monitored for the selected subset, and feature selection approaches that are embedded in adaptive systems, such as neural training algorithms.

The most common wrapper feature selection approach in remote sensing consists in analyzing the selection criteria for all the possible number of selected bands and most band combinations. For example, recursive feature elimination [Archibald and Fann, 2007] or genetic algorithms [Bazi and Melgani, 2006] optimizing the accuracy of support vector machines (SVM) have been successfully used in remote sensing data analysis. The main problem of these methods is their excessive computational cost, particularly high in the case of hyperspectral images, which often require including heuristics in the procedure. Besides, these feature selection methods may suffer from overfitting when working with a small number of training samples, as it is typically the case in remote sensing data processing.

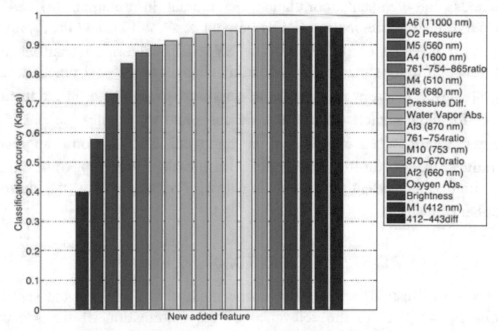

Figure 3.1: Classification accuracy (kappa statistic) of the multilayer perceptron (MLP) for the first 18 sets of features selected by the Sequential Forward Selection (SFS) algorithm.

The sequential feature selection algorithms identify the most relevant bands (e.g., those that better discriminate among classes) in a two-stage iterative feature selection process. Firstly, a search strategy for feature group selection is carried out [Zhang et al., 2007], and secondly, the objective function that evaluates the different subgroups is calculated. For example, in classification problems, a classification accuracy measure, such as the estimated Cohen's kappa statistic [Congalton and Green, 1999], of the classifier trained with the spectral bands and features under study might be used as objective function. Among all the available sequential selection approaches, the Sequential Forward Selection (SFS) method [Serpico and Bruzzone, 2001] will produce a hierarchy of the input bands by selecting first the best individual band and then adding the next best band (given the already selected bands) for each dimension.

In [Camps-Valls et al., 2010a], a filter approach was successfully compared to several standard and state-of-the-art filter and wrapper methods: the Pearson's correlation, R; mutual information, MI; SVM recursive feature elimination (RFE) [Guyon et al., 2002], and the L_0-norm approach [Weston et al., 2003]. The SVM-RFE algorithm analyzes the relevance of input variables by estimating changes in the cost function, $\Delta J_u = \|\mathbf{w}\| - \|\mathbf{w}_u\|$, where $\mathbf{w}$ represents the classifier weight vector in the feature space for the complete set of input variables and $\mathbf{w}_u$ denotes the classifier weight vector when variable u is removed. The L_0-norm approach iteratively seeks for the maximum classification accuracy of a linear classifier while essentially restricting $\|\mathbf{w}\|_0 < r$, where r is the desired number of features.

3.2.3 FEATURE SELECTION EXAMPLE

In order to illustrate a feature selection method in a real remote sensing problem, we analyze the relevance of the information of the spectral channels and the extracted features for a cloud detection problem. In particular, we show results for the sequential forward selection (SFS) algorithm using as selection criteria the classification accuracy of a multilayer perceptron (MLP) artificial neural network, which has shown better classification results than classification trees or support vector machines in our problem (see Chapter 4 for more information on image classification methods).

In this section, data from two real spectrometers onboard an Earth observation satellite will be used to illustrate the performance of the analyzed methods. The Medium Resolution Imaging Spectrometer (MERIS) [Rast et al., 1999] and the Advanced Along Track Scanning Radiometer (AATSR) [Llewellyn-Jones et al., 2001] instruments on board the Environmental Satellite (ENVISAT) are used together since they provide similar spatial resolution and swath, encompassing different spectral domains and viewing geometries. Images acquired by both MERIS and AATSR instruments provide complementary information in the visible and near-infrared and in the visible and infrared ranges of the electromagnetic spectrum, respectively. The set of features analyzed consist of 15 MERIS bands, 7 nadir plus 7 forward AATSR bands, and 33 cloud features extracted from both sensors attending to the cloud physical properties. We will come back later in Section 3.4 for details on physically meaningful features.

Figure 3.1 shows the value of the objective function for each number of features, i.e., the classification accuracy (kappa statistic) of the NN for the first 18 sets of features selected by the SFS algorithm. In this plot, we can check monotonicity up to 15 features (no local minima when adding features) but also the low improvement from 12 to 15 features. In addition to numerical results, its physical interpretation is of paramount importance for this particular application. All bands are not equally relevant for cloud screening and the combination of thermal info, the O_2 atmospheric absorption, and two VIS and SWIR bands produces the most significant improvement (see Table 3.2 in Section 3.4 for more details).

3.3 FEATURE EXTRACTION

Regarding the information contained in remote sensing images, many potential variables may be used in Earth observation applications, starting with the spectral signatures, passing through spectral indices or contextual information, and ending with linearly or nonlinearly transformed images. In this section, we review several approaches and methodologies that may be used for feature extraction in order to reduce the spectral and spatial data redundancy present in remote sensing data or to extract specific information related to the analyzed problem. The point now is that all input features can be eventually used to create the subset of new features, thus useful information may be fully retained at the expense of acquiring all input bands at the highest spatial resolution.

3.3.1 LINEAR METHODS

The use of linear projection methods, such as principal component analysis (PCA) [Jollife, 1986], is quite common in remote sensing data analysis. However, PCA can be quite limited for several reasons: the strong assumptions of linearity and Gaussianity, the fact that PCA only focuses on the directions of highest variance and not on any information criterion or class membership, or the issue of the noise characteristics.

Alternative methods to deal with these problems of PCA have been presented. For instance, the minimum noise fraction (MNF) transform [Green et al., 1988b] or the related noise-adjusted principal components (NAPC) include information about the noise features [Blackwell, 2005]. Also, orthogonal subspace projection (OSP) approaches have been proposed for hyperspectral image classification and dimensionality reduction [Harsanyi and Chang, 1994]. The partial least squares (PLS) approach [Wold, 1966] includes the information of the target variable (or labels) in the projection matrix, which is very convenient in classification problems, where the class labels are available: this leads to find a subspace containing the significant structure for this classification problem.

Other approaches including class-membership information have been presented. On the one hand, discriminant analysis feature extraction (DAFE) enhances class separability by means of maximizing the separation between their means, while minimizing the individual spread of the classes [Fukunaga, 1990]. On the other hand, decision boundary feature extraction (DBFE) extracts discriminant information from the decision boundary by looking for projections normal to the decision boundary [Lee and Landgrebe, 1993]. Other works have also presented semisupervised dimensionality reduction approaches for hyperspectral image classification

[Chen and Zhang, 2011]. In [Kuo and Landgrebe, 2004], a nonparametric weighted feature extraction (NWFE) was proposed to avoid numerical problems by computing nonparametric scatter matrices and weighting more the samples near the expected decision boundary.

Independent component analysis (ICA) [Comon, 1994] has been used in remote sensing to extract information contained in higher-order cross-moments of multivariate data [Zhang and Chen, 2002]; for example, to spatially and temporally deconvolve image sequences assuming the mutual statistical independence of the non-Gaussian physical sources [Lotsch et al., 2003]. ICA has been extensively used for the analysis of hyperspectral images [Wang and Chang, 2006b]; however, its use for unmixing hyperspectral data has been critically discussed in [Nascimento and Bioucas-Dias, 2005b] since the assumption of statistically independent sources does not apply for constrained mixtures. In this respect, also clustering algorithms have been used for dimensionality reduction via spectral feature extraction [Mojaradi et al., 2009]. Finally, also wavelet transforms has been proposed for feature extraction in the spectral domain to reduce the dimensionality of hyperspectral data [Bruce et al., 2002]. This use comes from the similarity between ICA and wavelet basis functions of remote sensing images, or the similarity between DCT and PCA, as shown in Chapter 2.

3.3.2 NONLINEAR METHODS

All previous methods assume that there exists a second order *linear* relation between the original input and target data matrices and the extracted input and output projections. However, in many situations this linearity assumption is not satisfied, and nonlinear feature extraction is needed to obtain acceptable performance.

Although nonlinear versions of algorithms such as PCA [Kramer, 1991] and ICA [Burel, 1992, Hyvärinen and Pajunen, 1999] have been presented in the literature, these algorithms have been scarcely used in remote sensing [Del Frate and Schiavon, 1999, Laparra et al., 2009, Licciardi et al., 2009a].

Other dimensionality reduction approaches are based on local metric information to represent the nonlinear structure of data [Saul et al., 2006]. These algorithms take into account the underlying global geometry of the data set to derive a manifold coordinate system [Ham et al., 2004]. Isometric mapping (ISOMAP) was proposed in [Tenenbaum et al., 2000] and adapted in [Bachmann et al., 2005, 2006] to achieve full-scene global manifold coordinates exploiting the nonlinear structure of hyperspectral imagery. This approach relies on graph methods to derive geodesic distances on the high-dimensional hyperspectral data manifold. Locally linear embedding (LLE) was proposed in [Roweis and Saul, 2000] to map input features into a single global coordinate system of lower dimensionality. In [Mohan et al., 2007], LLE was modified to include spatial coherence by comparing pixels using their local surrounding structure in the image domain, and applied to hyperspectral image segmentation. Similar approaches has been presented to learn the global structure of nonlinear manifolds in remote sensing [Dianat and Kasaei, 2010, Ma et al., 2010].

In this context, recent advances to cope with nonlinearities in the data are based on multivariate kernel machines, which have been presented and successfully applied to remote sensing data. *Kernel methods* are a promising approach, as they constitute a nice framework to formulate nonlinear versions from linear algorithms [Schölkopf and Smola, 2002, Shawe-Taylor and Cristianini, 2004]. Since the early use of support vector machines in remote sensing [Camps-Valls and Bruzzone, 2005], several kernel-based feature extraction methods have been proposed in the field. A set of multivariate kernel feature extraction methods, such as kernel PCA (KPCA), kernel PLS (KPLS), and kernel orthonormalized PLS (KOPLS), were proposed as a preprocessing step for hyperspectral image classification and canopy parameter retrieval [Arenas-García and Camps-Valls, 2008, Gómez-Chova et al., 2011b]. In [Gu et al., 2008], KPCA was also used for target and anomaly detection, while the kernel nonparametric weighted feature extraction (KNWFE) was introduced for classification in [Kuo et al., 2009]. Recently, a kernel version of the maximum autocorrelation function (MAF) has been successfully presented for change detection [Nielsen, 2011]. The explicit version of the MNF in reproducing kernel Hilbert spaces was proposed in [Gómez-Chova et al., 2011c].

Finally, the kernel entropy component analysis (KECA) nonlinear transform, which is based on information theory and does not have a linear counterpart, was applied for remote sensing image clustering [Gómez-Chova et al., 2011a]. Table 3.1 summarizes the main characteristics of both linear and kernel feature extraction methods.

Table 3.1: Summary of linear and nonlinear feature extraction methods based on multivariate analysis [Arenas-García and Petersen, 2009]. Notation: $\mathbf{X}$ and $\mathbf{Y}$ are the input and target data matrices, and $\tilde{\mathbf{X}}$ and $\tilde{\mathbf{Y}}$ their centered versions; $\mathbf{U}$ is the projection matrix that, for kernel methods, is expressed as a linear combination of the mapped samples $\mathbf{U} = \tilde{\mathbf{\Phi}}^{\top} \mathbf{A}$; $\mathbf{I}$ is a diagonal matrices of ones; $\mathbf{C}_{xx} = \frac{1}{n}\tilde{\mathbf{X}}^{\top}\tilde{\mathbf{X}}$ is the covariance matrix between all observations (input data); $\mathbf{C}_{xy} = \frac{1}{n}\tilde{\mathbf{X}}^{\top}\tilde{\mathbf{Y}}$ the covariance between the input and output data; $\mathbf{C}_{nn}$ is the estimated noise covariance matrix, $\mathbf{K}_{xx} = \tilde{\mathbf{\Phi}}^{\top}\tilde{\mathbf{\Phi}}$ is the kernel matrix between all observations; and $\mathbf{K}_{xn}$ between observations and the estimated noise.

Method	Maximized Function	Constraints	Solved Problem	Features
PCA	$\mathrm{Tr}\{\mathbf{U}^{\top}\mathbf{C}_{xx}\mathbf{U}\}$	$\mathbf{U}^{\top}\mathbf{U}=\mathbf{I}$	$\mathbf{C}_{xx}\mathbf{U}=\mathbf{U}\Lambda$	$\tilde{\mathbf{X}}\mathbf{U}$
	PCA projects linearly the input data onto the directions of largest input variance.			
MNF	$\mathrm{Tr}\{(\mathbf{U}^{\top}\mathbf{C}_{xx}\mathbf{U})/(\mathbf{U}^{\top}\mathbf{C}_{nn}\mathbf{U})\}$	$\mathbf{U}^{\top}\mathbf{C}_{nn}\mathbf{U}=\mathbf{I}$	$\mathbf{C}_{xx}\mathbf{U}=\mathbf{C}_{nn}\mathbf{U}\Lambda$	$\tilde{\mathbf{X}}\mathbf{U}$
	MNF maximizes the ratio between the signal and the noise variances for all the features.			
PLS	$\mathrm{Tr}\{\mathbf{U}^{\top}\mathbf{C}_{xy}\mathbf{V}\}$	$\mathbf{U}^{\top}\mathbf{U}=\mathbf{V}^{\top}\mathbf{V}=\mathbf{I}$	$\mathbf{C}_{xy}=\mathbf{U}\Lambda\mathbf{V}^{\top}$	$\tilde{\mathbf{X}}\mathbf{U}$
	PLS finds directions of maximum covariance between the projected input and desired output $\mathbf{Y}$.			
OPLS	$\mathrm{Tr}\{\mathbf{U}^{\top}\mathbf{C}_{xy}\mathbf{C}_{xy}^{\top}\mathbf{U}\}$	$\mathbf{U}^{\top}\mathbf{C}_{xx}\mathbf{U}=\mathbf{I}$	$\mathbf{C}_{xy}\mathbf{C}_{xy}^{\top}\mathbf{U}=\mathbf{C}_{xx}\mathbf{U}\Lambda$	$\tilde{\mathbf{X}}\mathbf{U}$
	OPLS finds optimal directions for performing linear regression of $\hat{\mathbf{Y}}$ on the projected input data.			
CCA	$\mathrm{Tr}\{\mathbf{U}^{\top}\mathbf{C}_{xy}\mathbf{V}\}$	$\mathbf{U}^{\top}\mathbf{C}_{xx}\mathbf{U}=\mathbf{I}\,\mathbf{V}^{\top}\mathbf{C}_{yy}\mathbf{V}=\mathbf{I}$	$\begin{pmatrix}\mathbf{0} & \mathbf{C}_{xy} \\ \mathbf{C}_{xy}^{\top} & \mathbf{0}\end{pmatrix}\begin{pmatrix}\mathbf{U}\\\mathbf{V}\end{pmatrix}=$ $\begin{pmatrix}\mathbf{C}_{xx} & \mathbf{0} \\ \mathbf{0} & \mathbf{C}_{yy}\end{pmatrix}\begin{pmatrix}\mathbf{U}\\\mathbf{V}\end{pmatrix}\Lambda$	$\tilde{\mathbf{X}}\mathbf{U}$
	CCA finds projections of the input and output data matrices such that each column of $\hat{\mathbf{Y}}$ can be reconstructed from the corresponding column of $\tilde{\mathbf{X}}$ with minimum square loss.			
KPCA	$\mathrm{Tr}\{\mathbf{A}^{\top}\tilde{\mathbf{K}}_{xx}\tilde{\mathbf{K}}_{xx}\mathbf{A}\}$	$\mathbf{A}^{\top}\tilde{\mathbf{K}}_{xx}\mathbf{A}=\mathbf{I}$	$\tilde{\mathbf{K}}_{xx}\mathbf{A}=\mathbf{A}\Lambda$	$\tilde{\mathbf{K}}_{xx}\mathbf{A}$
	KPCA finds directions of maximum variance of the input data in $\mathcal{H}$.			
KMNF	$\mathrm{Tr}\{(\mathbf{A}^{\top}\tilde{\mathbf{K}}_{xx}^2\mathbf{A})/(\mathbf{A}^{\top}\tilde{\mathbf{K}}_{xn}\tilde{\mathbf{K}}_{nx}\mathbf{A})\}$	$\mathbf{A}^{\top}\tilde{\mathbf{K}}_{xn}\tilde{\mathbf{K}}_{nx}\mathbf{A}=\mathbf{I}$	$\tilde{\mathbf{K}}_{xx}^2\mathbf{A}=\tilde{\mathbf{K}}_{xn}\tilde{\mathbf{K}}_{xn}^{\top}\mathbf{A}\Lambda$	$\tilde{\mathbf{K}}_{xx}\mathbf{A}$
	KMNF finds directions of maximum signal to noise ratio of the projected data in $\mathcal{H}$.			
KPLS	$\mathrm{Tr}\{\mathbf{A}^{\top}\tilde{\mathbf{K}}_{xx}\tilde{\mathbf{Y}}\mathbf{V}\}$	$\mathbf{A}^{\top}\tilde{\mathbf{K}}_{xx}\mathbf{A}=\mathbf{V}^{\top}\mathbf{V}=\mathbf{I}$	$\tilde{\mathbf{K}}_{xx}\hat{\mathbf{Y}}=\mathbf{A}\Lambda\mathbf{V}^{\top}$	$\tilde{\mathbf{K}}_{xx}\mathbf{A}$
	KPLS finds directions of maximum covariance between the input data in $\mathcal{H}$ and the output $\mathbf{Y}$.			
KOPLS	$\mathrm{Tr}\{\mathbf{A}^{\top}\mathbf{K}_{xx}\mathbf{K}_{yy}\mathbf{K}_{xx}\mathbf{A}\}$	$\mathbf{A}^{\top}\mathbf{K}_{xx}^2\mathbf{A}=\mathbf{I}$	$\mathbf{K}_{yy}\mathbf{K}_{xx}\mathbf{A}=\mathbf{K}_{xx}\mathbf{A}\Lambda$	$\tilde{\mathbf{K}}_{xx}\mathbf{A}$
	KOPLS extracts features that minimize the residuals of a multiregression approximating $\hat{\mathbf{Y}}$.			
KCCA	$\mathrm{Tr}\{\mathbf{A}^{\top}\mathbf{K}_{xx}\tilde{\mathbf{Y}}\mathbf{V}\}$	$\mathbf{A}^{\top}\mathbf{K}_{xx}^2\mathbf{A}=\mathbf{I}\,\mathbf{V}^{\top}\mathbf{C}_{yy}\mathbf{V}=\mathbf{I}$	$\begin{pmatrix}\mathbf{0} & \mathbf{K}_{xx}\hat{\mathbf{Y}} \\ \hat{\mathbf{Y}}^{\top}\mathbf{K}_{xx} & \mathbf{0}\end{pmatrix}\begin{pmatrix}\mathbf{A}\\\mathbf{V}\end{pmatrix}=$ $\begin{pmatrix}\mathbf{K}_{xx}^2 & \mathbf{0} \\ \mathbf{0} & \mathbf{C}_{yy}\end{pmatrix}\begin{pmatrix}\mathbf{A}\\\mathbf{V}\end{pmatrix}\Lambda$	$\tilde{\mathbf{K}}_{xx}\mathbf{A}$
	KCCA finds directions in feature space of maximum correlation with a certain projection of $\hat{\mathbf{Y}}$.			

3.3.3 FEATURE EXTRACTION EXAMPLES

In order to illustrate the performance of linear and nonlinear feature extraction methods, two examples are presented. The first one compares several linear and nonlinear feature extractions for posterior image classification. The second one presents an unsupervised feature extraction problem where the objective is to reduce the noise contribution in the features extracted from a hyperspectral image.

Remote Sensing Data

For both experiments we used the standard AVIRIS image taken over NW Indiana's Indian Pine test site in June 1992. The high number of narrow spectral bands (220 channels) induce a high collinearity among them. Discriminating among the major crops in the area can be very difficult (in particular, given the moderate spatial resolution of 20 meters), which has made the scene a challenging benchmark to validate classification accuracy of hyperspectral imaging algorithms. The calibrated data is available online (along with detailed ground-truth information) from `http://dynamo.ecn.purdue.edu/biehl/MultiSpec`. In all our experiments we used the whole scene, consisting of the full 145 × 145 pixels, which contains 16 classes, ranging in size from 20 − 2468 pixels, and thus constituting a very difficult situation. Among the 10366 pixels for which labels are available, 20% of the data is used for training the feature extractors, keeping the remaining 80% to test the discriminative power of the extracted features.

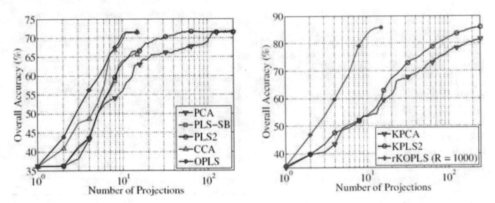

Figure 3.2: Average kappa statistic as a function of the number of projections extracted by several feature extraction methods. Left: PCA, PLS, PLS2, CCA, and OPLS. Right: KPCA, KPLS2, and the sparse KOPLS method (rKOPLS) with $R = 1000$. Figures adapted from Chapter 14 in [Camps-Valls and Bruzzone, 2009].

Experiment 1: linear and nonlinear feature extraction

In this experiment, we removed 20 noisy bands covering the region of water absorption, and finally worked with 200 spectral bands. Figure 3.2 shows the estimated kappa statistic in classification when using the features extracted with several linear and nonlinear methods for a varying number of features n_p. For the linear feature extraction, OPLS outperformed all other methods, closely followed by CCA. However, note that OPLS projections are easier to compute, since CCA comes as the solution to a more complex and ill-posed generalized eigenvalue problem. When the maximum number of projections are used, all methods but PLS-SB achieve the same error, but PCA and PLS2 require 200 features (i.e., the dimensionality of the input space), while 15 projections suffice for CCA and OPLS. For the non-linear case, we considered KPCA, KPLS2 and KOPLS with Gaussian kernel $k(\mathbf{x}_i, \mathbf{x}_j) = \exp(-\|\mathbf{x}_i - \mathbf{x}_j\|_2^2/2\sigma^2)$, using a 10-fold cross-validation procedure on the training set to select σ. Figure 3.2(right) displays the classification accuracy for the three methods. The sparse rKOPLS method (with $R = 1000$) was used instead of the more complex standard KOPLS. The same conclusions obtained for the linear case apply also to these nonlinear methods: rKOPLS achieves a superior performance than the other two methods, which require a much larger number of projections to achieve comparable performance. KPLS2 also outperforms KPCA, confirming the convenience of using the target matrix when feature extraction is considered for supervised learning. In the limit of n_p, the behavior of the three methods is very similar. However, we can obtain much better results from rKOPLS with a lower computational and memory burden. In general, non-linear methods obtain much higher recognition rates than linear techniques (72 % vs 85 %, approximately).

Figure 3.3: Extracted features from the AVIRIS image. From top to bottom: PCA, MNF, KPCA, and KMNF in the kernel space for the first 18 principal components. From left to right: each subimage shows the RGB composite of 3 components ordered in descending importance.

Experiment 2: signal-to-noise linear and nonlinear feature extraction

The second example is devoted to the application of different feature extraction approaches based on the maximization of the signal variance and the signal-to-noise ratio through linear and nonlinear transforms. In this example, all the 220 original bands are transformed onto a lower dimensional space. It is worth noting that 20 bands covering the region of water absorption are really noisy, thus allowing us to analyze the robustness of the different feature extraction methods to real noise. The significance of the results is tested for the different methods through the visual inspection of the extracted features and the performance obtained when using the extracted features in the same image dataset as before.

Visual inspection of the extracted features in descending order of relevance (Fig. 3.3) shows that the proposed approach provides more noise-free features than the other methods [Gómez-Chova et al., 2011c]. In addition, when using these features for classifying the land covers in the image, the classification accuracy increases as the number of used extracted features. Figure 3.4(a) shows how the best feature extraction methods are the linear MNF and the explicit KMNF. Finally, results are analyzed by inspecting the classification maps obtained with a linear discriminant analysis (LDA) classifier using the best sets of (linearly and nonlinearly) extracted features. The KMNF approach provides more spatially homogeneous land cover maps than the other methods (Fig. 3.4(b)).

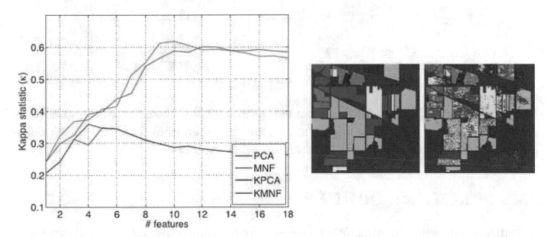

Figure 3.4: (a) Classification accuracy (kappa) as a function of the number of used features extracted with different methods. (b) On the top, RGB composite of the hyperspectral image (bright bare soils and dark vegetated crops) and *ground truth* of the 16 land-cover classes; On the bottom, classification maps using the MNF and KMNF features, respectively.

3.4 PHYSICALLY BASED SPECTRAL FEATURES

Previous sections have presented generic feature selection and extraction methods that are usually applied to remote sensing data. However, in remote sensing, one has a precise knowledge about how the signal is formed and acquired. The measured spectral signal at the sensor depends on the illumination, the atmosphere, and the surface, as explained in Chapter 1. Therefore, several physically-inspired features can be extracted from the spectrum before applying a machine learning algorithm: this typically improves its performance. Another possibility considers adapting standard feature extraction methods, such as PCA, to include knowledge about the physical problem [Bell and Baranoski, 2004]. As an ilustrative example, Fig. 3.5 shows the spectral curve of healthy vegetation, bare soil, and the atmospheric transmittance which are compared with the spectral channels of MERIS&AATSR spectrometers. The spectral bands free from atmospheric absorptions contain information about the surface reflectance, while others are mainly affected by the atmosphere.

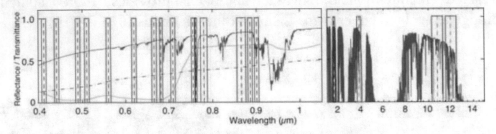

Figure 3.5: MERIS and AATSR channel locations (red boxes) superimposed to a reflectance spectra of healthy vegetation (green thin solid line), bare soil (black dash-dotted line), and the atmospheric transmittance (blue solid line).

3.4.1 SPECTRAL INDICES

Spectral indices are simple parametric combinations of several spectral channels. They are designed to reinforce the sensitivity to particular bio-physical phenomena, such as greenness, water content, etc. See Chapter 6 for a deeper description of the estimation of physical parameters from remote sensing images. Also, other factors affecting the shape of the spectral signature are taken into account in the design of the indices, such as soil contribution, solar illumination, atmospheric conditions, and sensor viewing geometry. A good spectral feature should not be sensitive to these factors. Many of these indices are extensively used in remote sensing because they rely on simple physical properties, and thus can be derived indistinctly from both multispectral or hyperspectral sensors. Therefore, the simple calculation of the indices has made possible deriving reasonable maps of bio-physical properties in a quick and easy way. Nevertheless, the majority of the indices only use up to five bands. Their simplicity proved to be desirable for numerous mapping applications, but it is also recognized that they under-exploit the full potential of the hyperspectral data cube [Schaepman et al., 2009].

3.4.2 SPECTRAL FEATURE EXTRACTION EXAMPLES

The first example presents one of the most widely used indices in the literature based on multispectral band ratios and arithmetic transforms: the Normalized Difference Vegetation Index (NDVI) [Liang, 2004, Rouse et al., 1973, Tucker, 1979], which is defined as NDVI = (NIR − R)/(NIR + R). The idea underlying this expression is that the difference between the visible red (R) and near-infrared (NIR) bands should be large for vegetated land covers due to the high chlorophyll absorption in the red and the high reflectance of the leaves in the near infrared (see green curve in Fig. 3.5). The normalization simply discounts the effects of uneven illumination.

In particular, NDVI is widely used because it is very general and simple to compute with only two bands that can be found in most imaging systems and even in airborne cameras. Similarly, many indices are based on band ratios, differential absorptions, or the area under the reflectance curve in some specific spectral regions. However, these ratio-based indices do not exploit the wealth of spectral information in other bands of multispectral and hyperspectral sensors. These families of physically-based ratios will be studied in detail in Chapter 6. Figure 3.6 shows an airborne orthoimage acquired with a spatial resolution of 0.5m using a camera with four channels (RGB and NIR). The scene presents an almond-tree plantation where tree crown can be easily highlighted in this image by computing the NDVI or the inverse of the intensity of the multispectral image.

Table 3.2: Cloud features extracted from MERIS and AATSR products (pixel-based).

Sensor	Cloud Feature	Channels Involved	Reference
MERIS	Brightness & Whiteness (VIS)	VIS bands [1-8]	Gómez-Chova et al. [2007]
MERIS	Brightness & Whiteness (NIR)	NIR bands [9 10 12 13 14]	Gómez-Chova et al. [2007]
MERIS	Brightness & Whiteness	VIS&NIR bands (without 11 & 15)	Gómez-Chova et al. [2007]
MERIS	O_2 absorption	754, 761, 778 nm	Gómez-Chova et al. [2007]
MERIS	WV absorption	885, 900nm	Gómez-Chova et al. [2007]
MERIS	Surface Pressure	761&754nm	Lindstrot et al. [2009]
MERIS	Surface Pressure	761/754nm ratio	European Space Agency [2006]
MERIS	Bright over Land (sand)	443/754nm ratio	European Space Agency [2006]
MERIS	Bright over Land (ice)	709/865nm ratio	European Space Agency [2006]
MERIS	Cirrus over Ocean/Land	761/754nm ratio ; 865nm	European Space Agency [2006]
MERIS	Bright Clouds	450nm	Preusker et al. [2008]
MERIS	Snow Test (reflectance)	865/890 NDI	Preusker et al. [2008]
MERIS	Cloud 412 reflectance	412/443nm ratio	Kokhanovsky et al. [2008]
MERIS	Cloud 412 reflectance	412/443nm difference	Kokhanovsky et al. [2008]
MERIS	Cloud mask 1	412/681nm ratio	Guanter et al. [2008]
MERIS	Cloud mask 2	412/708nm ratio	Guanter et al. [2008]
MERIS	Hue-Saturation-Value transf.	665, 560, 442nm	González and Woods [2007]
AATSR	Gross Cloud	$12\mu m$	European Space Agency [2007]
AATSR	Thin Cirrus	$11/12\mu m$ difference	European Space Agency [2007]
AATSR	$11/12\mu m$ Nadir/Forward	$11\mu m$ nad/fwd ; $11/12\mu m$	European Space Agency [2007]
AATSR	Visible Channel Cloud Test	870,670,550nm NDI	Prata [2002]
AATSR	Snow Test	$1.6\mu m$ 550nm NDI	Prata [2002]
AATSR	Reflectance Gross Cloud	670nm	Birks [2007]
AATSR	Reflectance Ratio	870/670nm ratio	Birks [2007]
AATSR	Albedo	$3.7\mu m$	Birks [2007]
AATSR	Thermal Difference	$11/12\mu m$ difference	Birks [2007]
AATSR	Thermal Gross Cloud	$11\mu m$	Birks [2007]
AATSR	$11\mu m$ Nadir/Forward	$11\mu m$ nad/fwd	Muller and Brinckmann [2008]
AATSR	865 Nadir/Forward	865 nad/fwd	Muller and Brinckmann [2008]

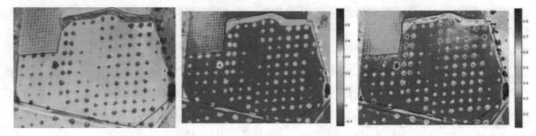

Figure 3.6: RGB composite (*left*), NDVI (*center*), and inverse of the intensity (*right*) features extracted from a 4-channel airborne orthoimage.

The second example, following with the example presented in Section 3.2.3, presents a list of features based on physical characteristics that are useful to discriminate clouds from surface (Table 3.2). The proposed cloud-features are extracted from the MERIS and AATSR spectral bands. We indicate for each feature the original bands used to compute them and references with further information. Note that all the listed features are extracted from the information contained in a given pixel, i.e., no spatial or contextual information is used, which will be presented in the following sections.

3.5 SPATIAL AND CONTEXTUAL FEATURES

Spatial resolution of current remote sensing instruments allows very fine scale image processing. The analysis of objects and contextual information in very high resolution remote sensing images usually provides a more meaningful interpretation of the data. This image content analysis through the recognition of objects, as opposed to a pixel based approach, is closer to human perception and presents strong relations with computer vision, image analysis and pattern recognition [Argialas and Harlow, 1990].

Spatial filtering of remote sensing images is a crucial issue to avoid noise or objects which are undesirable for the image processing purpose, while maintaining the detail provided by the images. This can be achieved by adding contextual features to the spectral information. See Chapter 4.2.3 for a land-cover classification example with and without contextual information. In remote sensing, contextual information has been mainly considered with different strategies such as *convolution filters* [Clausi and Deng, 2005, Kandaswamy et al., 2005, Kruizinga et al., 1999], *textural features* and *Markov random fields* [Clausi and Yue, 2004, Lorette et al., 2000, Rellier et al., 2004, Zhao et al., 2007], *mathematical morphology* [Fauvel et al., 2008, Pesaresi and Benediktsson, 2001], or *spatial transforms* [Li et al., 2011, Shah et al., 2010].

3.5.1 CONVOLUTION FILTERS

These kinds of filters perform the convolution product of a specifically designed two-dimensional spatial impulse response with an image. The convolution operation [González and Woods, 2007] places the filter matrix centered on the original image pixel and provides the output value as the sum of the product, element by element, of the filter matrix by the original image patch covered by the filter. Therefore, the convolution modifies the gray level values of the input image according to the filter applied to it.

The operation of convolution is used to highlight features of interest as objects, edges and contours [Ehrich, 1977] or for image enhancement such as noise removal or smoothing [Womack and Cruz, 1994]. Usually, banks of convolution filters are used for feature extraction. For instance, filter banks consisting of a number of Gabor filters with different frequencies, resolutions, and orientations are commonly used to extract textural features in remote sensing images [Augusteijn et al., 1995, Kruizinga et al., 1999].

3.5.2 CO-OCCURRENCE TEXTURAL FEATURES

Texture analysis and classification is a critical aspect in image processing. Among the methods developed to model natural scenes (e.g., moment-based, wavelet transforms, Gabor filters, Markov random fields, etc.), an efficient statistical method to model image textures is the construction of grey level co-occurrence matrices (GLCMs) [Haralick et al., 1973]. Co-occurrence matrices are probability-distribution matrices computed for a given pixel that try to capture the spatial variation as a function of the orientation and distance among neighboring pixels. GLCM features have been extensively used in remote sensing image processing from the very beginning [Baraldi and Parmiggiani, 1995] and compared with other classical approaches [Clausi and Deng, 2005]. However, their application to remote sensing data has shown clear advantages for synthetic aperture radar (SAR) images; in particular, for the analysis and classification of SAR sea ice imagery [Barber and Ledrew, 1991, Clausi, 2001, Clausi and Yue, 2004, Soh and Tsatsoulis, 1999].

3.5.3 MARKOV RANDOM FIELDS

Gaussian Markov random field (MRF) models [Cross and Jain, 1983, Hassner and Sklansky, 1980] are also an extended method for texture feature extraction. The idea behind MRF models is to characterize the statistical relationship between the intensity at a given pixel and its neighbors within a given neighborhood structure. The mean intensity of the pixel is modeled as a weighted average of the intensities of selected neighbors allowing a zero mean stationary Gaussian noise. For each sliding patch or window, the unknown weights of the model are estimated through least squares and then are used as the features describing the texture properties of the central pixel.

In remote sensing, these kind of textural features have been used for classification of high resolution images in urban areas [Lorette et al., 2000].

In [Rellier et al., 2004], MRF have been also used for the classification of hyperspectral images. In [Clausi and Yue, 2004], however, MRF were used for texture analysis of SAR sea ice imagery. Therefore, the main application of these texture features in remote sensing is mainly the classification of high spatial resolution imagery [Zhao et al., 2007].

3.5.4 MORPHOLOGICAL FILTERS

A way to integrate contextual information is to extract shape information of single objects using *mathematical morphology* [Serra, 1982, Soille, 2004]. Mathematical morphology provides a collection of image filters (called operators) based on set theory.

Morphological operators

Morphological operators are essentially a collection of filters based on set theory. Morphological operators [Serra, 1982, Soille, 2004] can be summarized into two fundamental operations: *erosion* and *dilation*. Basically, erosion deletes all pixels which neighborhood cannot contain a certain structuring element while dilation provides an expansion by addition of the pixels contained in the structuring element. Binary morphology has been extended to grayscale images by considering them as a topographic relief, where brighter tones correspond to higher elevation [Plaza et al., 2005, Sternberg, 1986].

Two of the most common morphological operators are the *opening* and the *closing* operators. Opening is the dilation of an eroded image and is widely used to filter brighter (compared to surrounding features) structures in grayscale images. On the contrary, closing is the erosion of a dilated image and allows one to filter out darker structures [Sternberg, 1986]. With classical opening and closing operators, although the size of the objects is properly captured, the shapes of the objects in the image are missed or distorted. In general, this is not a desirable behavior in image processing. In order to preserve original shapes in the morphological images, the use of *opening* (and *closing*) *by reconstruction* operators instead of simple opening and closing operators was proposed in [Crespo et al., 1995]. All these morphological operations are nowadays standard tools in remote sensing feature extraction, widely applied in image segmentation (see Chapter 4). Figure 3.7 represents an example of erosion, dilation, closing and opening morphological operators using a diamond-shaped structuring element.

Figure 3.7: On the left end the original intensity image and then an example of erosion, dilation, closing and opening morphological operators using a diamond-shaped structuring element.

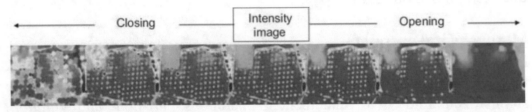

Figure 3.8: Progressive opening and closing using a diamond-shaped structuring element. On the left, the closing images produced using three structuring element of increasing size; in the middle, the original image; on the right end, the opening images.

Morphological profiles

To fully exploit the information about the shape of objects present in the scene at a pixel level, Pesaresi and Benediktsson [2001] proposed a multiscale analysis of morphological transforms to build a morphological profile. The morphological profile of a pixel consists of the values of a series of openings (and closings) obtained by applying structuring elements of increasing sizes (see an illustrative example in Fig. 3.8). This profile characterizes the shape of the homogeneous objects in the image and higher level information can be further extracted from it, such as the derivative of the morphological profile, which shows the residuals of two successive filtering operations (i.e., two adjacent opening and closing levels in the profile). In particular, the residuals of an opening (or a closing) image with respect the original image, which are called top-hat operators, are commonly used to extract this kind of information.

A review of morphology in remote sensing

In remote sensing, morphological operators have been used to classify remote sensed images at metric resolutions and have been highlighted as very promising tools for data analysis [Pesaresi and Benediktsson, 2001, Soille and Pesaresi, 2002]. The effective results with panchromatic imagery [Sternberg, 1986, Wang et al., 1994] using basic operators, such as opening and closing, have focused the attention of the scientific community on the use of morphological analysis. In [Epifanio and Soille, 2007, Pina and Barata, 2003], the authors use morphological operators for image segmentation. Pesaresi and Benediktsson proposed building differential morphological profiles (DMP) to account for differences in the values of the morphological profiles at different scales. Such profiles have been used in [Pesaresi and Benediktsson, 2001, Pesaresi and Kannellopoulos, 1999] for IRS-1C panchromatic image segmentation, using the maximal derivative of the DMP, because this value shows the optimal size of the structure of the object. In [Benediktsson et al., 2003], complete DMP have been used for both IRS-1C and IKONOS panchromatic image classification. In particular, reconstruction DMP have been used with a neural network classifier and two linear feature extraction methods have been proposed in order to reduce the redundancy in the information. More recently, the extraction of morphological information from hyperspectral imagery has been discussed.

In recent literature [Benediktsson et al., 2005, Fauvel et al., 2008, Palmason et al., 2003], the first principal component of the image has been used to extract the morphological images. In [Louverdis et al., 2002], morphological operators have been proposed for multichannel images. In [Plaza et al., 2003], extended opening and closing operators by reconstruction are used for target detection on both AVIRIS and ROSIS data, while in [Plaza et al., 2005] multichannel and directional morphological filters are proposed for hyperspectral image classification, showing the ability to account for scale and orientation of objects simultaneously. Recently, in [Tuia et al., 2009a], multichannel morphology is used for change detection of high resolution images using SVM. Finally, Dalla Mura et al. [2010] applied morphological attribute profiles for the analysis of very high resolution images, which consider more morphological attribute operators in addition to the standard openings and closings by reconstruction.

3.5.5 SPATIAL TRANSFORMS

In previous sections, we have seen how the smoothness of the signals measured by remote sensing instruments in the spectral domain can be used to reduce the dimensionality of the data. This smoothness or statistical redundancy also holds for the spatial domain in hyperspectral remote sensing images at very different spatial resolutions. This fact, as observed in Chapter 2, explains why the tools developed for conventional images are expected to be successful when applied to the analysis of hyperspectral images. Therefore, spatial smoothness justifies the use of transforms like PCA and local DCT. On the other hand, edges between spatially stationary regions in remote sensing images give rise to wavelet-like basis functions when linear ICA is applied to the spatial dimension for image feature extraction [Birjandi and Datcu, 2010].

Traditional image processing techniques have been applied to remote sensing data. For example, the two-dimensional Fourier transform has been used for image registration [Xu and Varshney, 2009] and denoising [Fattahi et al., 2009], but also for textural feature extraction. It is well known that the Fourier power spectrum is sensitive to texture coarseness. A coarse texture will have high values concentrated near the origin, while in a fine texture the values will be more spread out. Similarly, it is also known that the angular distribution of values of the Fourier power spectrum is sensitive to the directionality of the texture. In [Weszka et al., 1976], texture features based on the discrete Fourier power spectrum were used for pattern classification of optical images, while a Fourier-based textural feature extraction was proposed for radar images in [Stromberg and Farr, 1986]. The same spatial correlation among pixels, for a series of separation distances and directions, has been exploited by the experimental variogram, which computes the arithmetic average of the squared difference between pairs of values at a given distance [Journel and Huijbregts, 1978]. In remote sensing images, the experimental variogram is usually an exponential increasing function that saturates when the distance between samples increase and the values cease to be correlated.

These spatial correlations can be also efficiently exploited by linear techniques, such as PCA or ICA and local DCT or wavelets, to find features removing the statistical redundancy between the transform coefficients. As mentioned in Chapter 2, spatial PCA eigenfunctions obtained for remote sensing image patches are similar to the basis functions of the local discrete cosine transform. Analogously, the non-Gaussian nature of the remote sensing signals justifies the use of wavelet-based transforms, which are also similar to the basis functions found when applying ICA to remote sensing image patches. For example, in [Chen and Qian, 2011], a two-dimensional bivariate wavelet thresholding method is used to remove the noise for low-energy PCA channels, and a one-dimensional dual-tree complex wavelet transform denoising method is used to remove the noise of the spectrum of each pixel of the data cube. Also, in [Li et al., 2011], a two-dimensional oriented wavelet transform (OWT) is introduced for efficient remote-sensing image compression.

Wavelet transforms are localized in both space and frequency whereas the Fourier transform is only localized in frequency. In [Li and Narayanan, 2004], textural features characterizing spatial information are extracted using Gabor wavelet coefficients. In [Fattahi et al., 2009], windowed Fourier transform and wavelet transform are used to reduce synthetic aperture radar interferometric phase noise. Wavelets are also strongly related to scale-space transforms, and can be used for multi-scale image representation. In addition, in remote sensing, they are used to fuse multispectral and panchromatic images [González-Audicana et al., 2004]. In this context, wavelets may as well be used for spatial feature extraction in remote sensing images. In [Simhadri et al., 1998], a multiresolution decomposition is used for extracting the features of interest from the images by suppressing the noise. In [Fukuda and Hirosawa, 1999], a wavelet-based texture feature set is derived for land cover classification in SAR images. In [Meher et al., 2007], extracted features obtained by the wavelet transform, rather than the original multispectral features of remote sensing images, are used for land-cover classification. Finally, in [Shah et al., 2010], a feature set is obtained by combining ICA and the wavelet transform for image information mining in geospatial data.

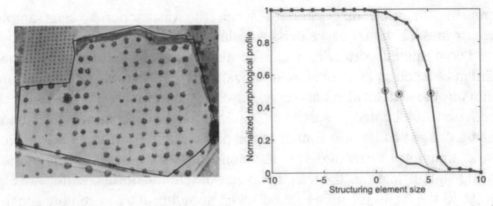

Figure 3.9: RGB composite of an almond grove (left) and normalized morphological profiles (right) corresponding to the centers (red points on the RGB image) of three almond-trees of different crown sizes.

It is worth noting that, as shown in Chapter 2, the smoothness of both the spectral and the spatial dimensions can be considered at the same time (see Fig. 2.4). This joint smoothness justifies the use of joint three-dimensional local frequency transforms in the spectral and the spatial dimensions for hyperspectral processing. In Section 2.3, benefits of using joint spatio-spectral transforms are shown for an efficient coding example. Although this kind of approach is not very extended in the remote sensing community, several techniques are available for a joint transform along the spatial and spectral axes. Renard and Bourennane [2009] represented hyperspectral images as tensors to jointly take advantage of the spatial and spectral information. They performed a dimensionality reduction on the spectral domain using PCA along with an orthogonal projection onto a lower subspace dimension of the spatial domain. Dianat and Kasaei [2010] proposed an approach that takes into account the spatial relation among neighboring image pixels while preserving all useful properties of PCA. Bau et al. [2010] presented a new model for spectral/spatial information based on 3-D Gabor filters for capturing specific orientation, scale, and wavelength-dependent properties of hyperspectral image data. The field of remote sensing image segmentation, that will be revised in the next chapter, has exploited most of these spatial processing techniques, either with spatial preprocessing of the image or with spatio-spectral approaches: see a taxonomical review of the techniques in [Camps-Valls et al., 2010b].

3.5.6 SPATIAL FEATURE EXTRACTION EXAMPLE

In this section, very high resolution airborne data (see Fig. 3.6) is used to illustrate different spatial feature extraction methods at a spatial resolution of 0.5 m. The final objective is to characterize the properties of different tree plantations.

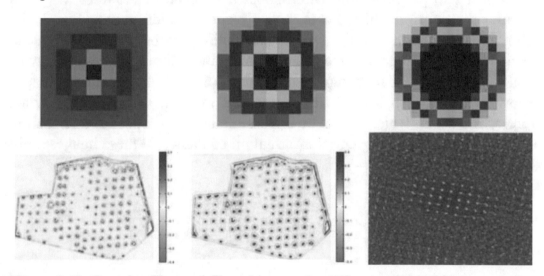

Figure 3.10: Gaussian filter and filtered images for different radius of the convolution kernel.

As explained in Section 3.5.4, the closing and opening operators (see Fig. 3.7) are used to build a morphological profile for each pixel (see Fig. 3.8). This profile characterizes the shape of the homogeneous objects in the image and higher level information can be further extracted from it. For instance, Fig. 3.9 shows the normalized morphological profiles corresponding to the centers (red points on the RGB image) of three almond-trees of different sizes. In these profiles, one can easily observe how the transition from high to low values of the structuring element size. The radius of the tree crown can be easily estimated from the morphological profile.

Once the size of the objects in the image has been estimated for each pixel, specific convolution filters can be used to find the trees present in the image. Figure 3.10 shows three radial basis functions of different sizes that are used to highlight trees from the background (bright circular objects) in the filtered image. One can see how, depending on the radius of the filter, trees of a size approximately equal to the filter radius are distinguished from the background.

The location of the trees can be further processed to characterize the plantation grid. The energy and frequency of the two-dimensional Fourier transform will provide information about the spread of the trees, while its angular distribution will indicate their direction (middle row of Fig. 3.11). Similar information about the trees distribution can be obtained from the experimental variogram in the principal plantation direction (bottom row of Fig. 3.11). The spatial features extracted form the high-resolution images allow to distinguish several tree plantations (almond, olive, orange, vine, forest, and bare), which would be impossible using the spectral information only due to the limited spectral information contained in these images: only RGB and NIR features.

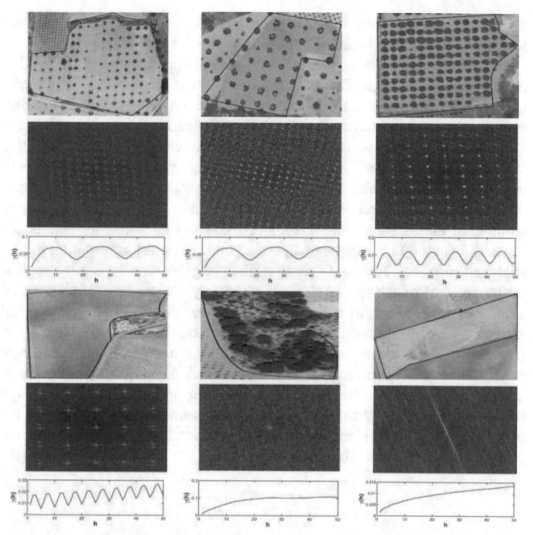

Figure 3.11: On the top, RGB composites of several tree plantations: almond, olive, orange, vine, forest, and bare). On the middle, two-dimensional Fourier transform of the trees within the red boundary. On the bottom, experimental variogram in the principal plantation direction.

3.6 SUMMARY

In this chapter, we analyzed which are the most common feature selection and feature extraction methods in remote sensing image processing. Due to the high spectral and spatial resolution of current remote sensing instruments, it is specially relevant to reduce the redundant information before further processing the images or to extract the important information for the specific remote sensing application.

Two different, although intimately related, approaches are followed to reduce the dimensionality of the data while preserving the useful information. First, we described feature selection, where those variables that do not positively contribute to the predefined task are identified and ignored. Despite the loss of information, feature selection is really attractive in remote sensing since it allows physical interpretation, which is very useful to transfer results from one sensor to another. Then, we continued with feature extraction, where a subset of new features is created by combining the existing ones. The objective is to find a transformation to a lower dimension feature space while retaining the useful information for the analyzed problem. In remote sensing, feature extraction is usually applied to the spectral or the spatial domain separately. On the one hand, spectral feature extraction performs linear and nonlinear transforms of the spectral signature that can be based on physical knowledge about the problem or on the statistical properties of the data. On the other hand, spatial feature extraction can be performed using a variety of techniques some of them inherited from image processing and other specifically designed to cope with remote sensing problems. Finally, we provided some hints about the importance of further exploring spatio-temporal transforms in remote sensing based on results from Chapter 2.

CHAPTER 4

Classification of Remote Sensing Images

The advances in remote sensing sensors (see Chapter 1) allow description of the objects of interest with improved spatial and spectral resolutions. The excellent quality in the acquired data gives rise at the same time to challenging problems for automatic image *classification*[1]. This chapter considers these challenges. First, we present different classification problems that benefit from the use of remote sensing data and briefly present the solutions considered in the field. After the definition of the classification problem (Section 4.1.1), the main classification tasks will be presented in sections 4.2 to 4.4. Finally, Section 4.5 discusses the new challenges to be confronted in these fields, including representativeness and adaptation of classifiers.

4.1 INTRODUCTION

Classification maps are one of the main products for which remote sensing images are used. The specific application may differ, but the general aim is to associate the acquired spectra to a limited number of classes. These classes are expected to facilitate the description or detection of objects at the Earth's surface. In the context of decision making, classification maps are useful, because they summarize the complex spatial-spectral information into a limited number of classes of interest.

4.1.1 THE CLASSIFICATION PROBLEM: DEFINITIONS

Consider a task such as distinguishing different soil types using their spectral signature, or discerning trees from buildings using aerial photography. All these problems share the task of predicting a class membership (or label) $y_i \in \mathbb{N}$ from a set of d-feature samples $\mathbf{x}_i = \{x_i^{(1)}, x_i^{(2)}, ..., x_i^{(d)}\}^\top \in \mathbb{R}^d$. The labels correspond to different classes, whose nature is specific to each applicative domain. The features are typically the spectral bands of a satellite/airborne image (for instance, $\mathbf{x} = \{x^{(1)} = $ blue band, $x^{(2)} = $ green band, $x^{(3)} = $ red band, $x^{(4)} = $ near infrared band$\}^\top$), spatial or contextual features extracted over some specific channels, or full hyperspectral patches. See Chapters 2 and 3 for details about these families of features. The variables (or features hereafter) included in $\mathbf{x}$ are supposed to be discriminative for the classification task. Figure 4.1 shows a generic flowchart of remote sensing image classification. Given the set of d features, two main schemes are adopted:

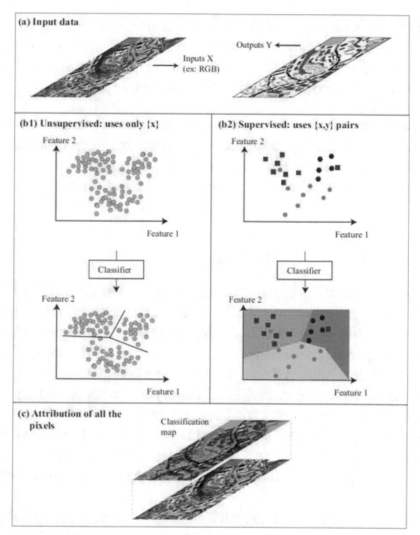

Figure 4.1: General remote sensing image classification problem.

- *Unsupervised* classification (Fig. 4.1(b1)), where the features are used to identify coherent clusters in the data distribution. The aim of such methods is to split the data into groups as similar as possible, without knowing the nature of such groups.

- *Supervised* classification (Fig. 4.1(b2)) refers to the situation in which a series of input-output pairs (the training set) are available and used to build a model. The model learns a function to assign output labels (or targets) to every input feature vector using the training set. In most settings, training data are assumed to be representative of the whole data distribution and to be independent and identically distributed (i.i.d.). We will come back about this statement in Section 4.5.5.

In both cases, a model with Θ parameters $\hat{y} = f(\mathbf{x}, \Theta)$ is constructed using the training data $\mathbf{x}$ and then applied to the entire image, in order to have the class estimate $\hat{y}$ of each pixel in the image. The main difference for the application of one of these approaches resides in the availability of training labels.

4.1.2 DATASETS CONSIDERED

In all the experiments presented in this chapter, two QuickBird images of Brutisellen –a residential neighborhood of Zürich, Switzerland– have been used to assess the performance of the algorithms presented. The images have been acquired in August 2002 and October 2006, respectively, and have a spatial resolution of 2.4 m. They are coregistered, 4-dimensional (Near Infrared, Red, Green and Blue bands) and with an extent of 329 × 347 pixels. A total of 40762 (2002 image) and 28775 (2006 image) pixels were labeled by photointerpretation and belong to 9 land-use classes. Figure 4.2 illustrates the images and the labeled pixels available. For each application, the nature of the outputs has been defined in the corresponding section.

In addition, we evaluate the performance of semisupervised algorithms in an AVIRIS image acquired over the Kennedy Space Center (KSC), Florida, on 1996, with a total of 224 bands of 10 nm bandwidth with center wavelengths from 400-2500 nm. The data was acquired from an altitude of 20 km and has a spatial resolution of 18 m. After removing low SNR bands and water absorption bands, a total of 176 bands remains for the analysis. The dataset originally contained 13 classes representing the various land cover types of the environment, containing many different marsh subclasses that we decided to merge them in a fully representative marsh class. Thus, we finally used 10 classes: 'Water' (761), 'Mud flats' (243), 'Marsh' (898), 'Hardwood swamp' (105), 'Dark/broadleaf' (431), 'Slash pine' (520),

'CP/Oak' (404), 'CP Hammock' (419), 'Willow' (503), and 'Scrub' (927). The image can be downloaded from `http://www.csr.utexas.edu/`.

4.1.3 MEASURES OF ACCURACY

Classification accuracy is typically evaluated on a test set independent from the training set, in order to assess generalization capabilities of the models. Several standard quality measures in the remote sensing literature are commonly used.

Overall accuracy The *overall accuracy* (OA) is the number of pixels correctly classified divided by the total number of test pixels: this measure shows the ratio between correct labels and commission errors, i.e., the pixels attributed to a wrong class.

Kappa index The overall accuracy can be a misleading index, since several correct classifications may occur by chance. Very often the estimated *Cohen's Kappa statistic* κ is preferred instead [Foody, 2004]. This index is computed by comparing the number of pixels classified correctly by the model to the number of pixels classified correctly by random chance.

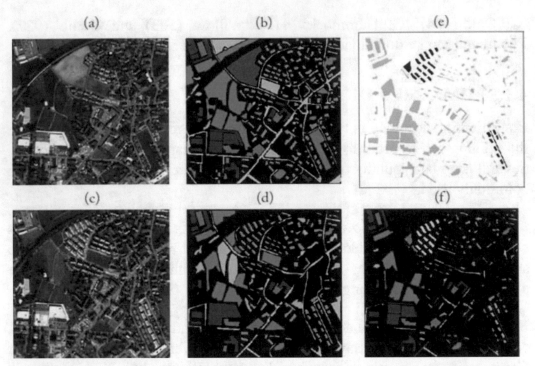

Figure 4.2: Images used in the experiments: Images of Brutisellen (Zürich, Switzerland) and corresponding ground surveys. Subfigures (a-b) show the 2002 image, while subfigures (c-d) show the 2006 image. Subfigures (e) and (f) illustrate the classes used for the change detection experiments: in panel (e), the changed and unchanged areas are highlighted in black and grey, respectively; panel (f) shows the color code used in Figure 4.4 (yellow = soil to buildings; cyan = meadows to building; light pink = soil to vegetation; dark pink = vegetation to road).

Mean tests In the case of multiple experiments reported in this survey, a mean test is used to see if the mean Kappa for the first model, $\bar{\kappa}_1$, is significantly higher than the mean kappa obtained by the second model, $\bar{\kappa}_2$. We thus use the standard formulation for a mean test statistic for two populations. We reject the hypothesis of $\bar{\kappa}_1$ being less or equal to $\bar{\kappa}_2$ if and only if

$$\frac{(\bar{\kappa}_1 - \bar{\kappa}_2)\sqrt{n_1 + n_2 - 2}}{\sqrt{(\frac{1}{n_1} + \frac{1}{n_2})(n_1 s_1^2 + n_2 s_2^2)}} > t_{1-\alpha}[n_1 + n_2 - 2], \tag{4.1}$$

where s_1 and s_2 are the observed standard deviations for the two models, n_1 and n_2 are the number of realizations of experiments reported and $t_{1-\alpha}$ is the α-th quantile of the Student's law (typically $\alpha = 0.05$ is used).

McNemar's test The McNemar's test [Foody, 2004], or Z-test, is a procedure used for assessing the statistical difference between two maps when the same samples are used for the assessment. It is a nonparametric test based on binary confusion matrices. The test is based on the z statistic:

$$z = \frac{f_{12} - f_{21}}{\sqrt{f_{12} - f_{21}}},$$ (4.2)

where f_{ij} indicates the number of pixels in the confusion matrix entry (i, j). This index can be easily adapted to multiclass results by considering correct (diagonal) and incorrect (off-diagonal) results. When z is positive (or, in accordance to 95-th quantile of the normal law $\mathcal{N}(0, 1)$, greater than 1.96), the result of the second map can be considered as statistically different. When $|z| < 1.96$, the maps cannot be considered as statistically significantly different, and when z is negative, the first result statistically outperforms the second one.

ROC curves In problems involving a single class, such as target detection (see Section 4.4), the *receiver operating characteristic* (ROC) curve is used to compare the sensitivity of models. By sensitivity, we mean the relation between the true positive rate (TPR) and the false positive rate (FPR). These two measures define the first and second type risk, respectively. In the experiments of Section 4.4, we use a traditional measure computed from the ROC curve, the Area Under the Curve (AUC): this area can be interpreted as the probability that a randomly selected positive instance will receive a higher score than a negative instance.

4.2 LAND-COVER MAPPING

Land-cover maps are the traditional outcome of remote sensing data processing. Identifying specific land cover classes on the acquired images may help in decision-making by governments and public institutions. The land-cover maps constitute an important piece to complement topographic maps, agriculture surveys, or city development plans, which are fundamental for decision-making and planning. Traditionally, these maps are retrieved by manual digitalization of aerial photography and by ground surveys. However, the process can be automatized by using statistical models as the ones illustrated in Fig. 4.1. Land-cover classification problems can be resumed to the task of classifying an image using information coming from this same image. The outputs represent a series of land-cover classes, for instance $y = \{1:$ vegetation, $2:$ man-made, $3:$ soil$\}$. In the following, the application of supervised and unsupervised methods to remote sensing images is briefly summarized (cf. Fig. 4.1).

4.2.1 SUPERVISED METHODS

The contribution of supervised methods has been improving the efficacy of the land-cover mapping methods since the 1970s: Gaussian models such as Linear Discriminant Analysis (LDA) were replaced in the 1990s by nonparametric models able to fit the distribution observed in data of increasing dimensionality. Decision trees [Friedl and Brodley, 1997, Hansen et al., 1996] and then more powerful neural networks (NN, Bischof and Leona [1998], Bischof et al. [1992], Bruzzone and Fernández Prieto [1999]) and support vector machines (SVM, Camps-Valls and Bruzzone [2005], Camps-Valls et al. [2004], Foody and Mathur [2004], Huang et al. [2002], Melgani and Bruzzone [2004]) were gradually introduced in the field, and quickly became standards for image classification. In the 2000s, the development of classification methods for land-cover mapping became a major field of research, where the increased computational power allowed 1) to introduce different types of information simultaneously and to develop classifiers relying simultaneously on spectral and spatial information [Benediktsson et al., 2003, 2005, Fauvel et al., 2008, Pacifici et al., 2009a, Pesaresi and Benediktsson, 2001, Tuia et al., 2009a], 2) the richness of hyperspectral imagery [Camps-Valls and Bruzzone, 2005, Plaza et al., 2009], and 3) exploiting the power of clusters of computers [Muñoz-Marí et al., 2009, Plaza et al., 2008].

Kernel methods were the most studied classifiers: ensembles of kernels machines [Briem et al., 2002, Waske et al., 2010], fusion of classification strategies [Bruzzone et al., 2004, Fauvel et al., 2006, Waske and J. A. Benediktsson, 2007] and the design of data-specific kernels –including combinations of kernels based on different sources of information [Camps-Valls et al., 2006b, Tuia et al., 2010a,c] or spectral weighting [Baofeng et al., 2008]– became major trends of research and attracted the increasing interest from users and practitioners.

4.2.2 UNSUPERVISED METHODS

Although less successful than supervised methods, unsupervised classification is still attracting a large consensus in remote sensing research. The problem of the acquisition of labeled examples makes unsupervised methods attractive and thus ensure constant developments in the field[2]. Two main approaches to unsupervised classification for land-cover mapping are found in the literature: *partitioning methods*, that are techniques that split the feature space into distinct regions, and *hierarchical methods*, that return a hierarchical description of the data in the form of a tree or *dendrogram*.

Partitioning methods are the most studied in remote sensing: first, fuzzy clustering has been used in conjunction with optimization algorithms to cluster land-cover regions in Maulik and Bandyopadhyay [2003] and Maulik and Saha [2009]. This approach is extended to multiobjective optimization in [Bandyopadhyay et al., 2007] where two opposing objective functions favoring global and local partitioning were used to enhance contextual regularization. Other regularization strategies include the fusion of multisource information [Bachmann et al., 2002, Sarkar et al., 2002] or rule-based clustering [Baraldi et al., 2006]. Graph cuts have been reconsidered more recently [Tyagi et al., 2008] where the authors proposed a multistage technique cascading two clustering techniques, graph-cuts and fuzzy c-means, to train the expectation-maximization (EM) algorithm. The use of self organizing maps is proposed in [Awad et al., 2007]. Tarabalka et al. [2009] use ISOMAP to regularize the SVM result by applying a majority vote between the supervised and the unsupervised method.

Hierarchical methods cluster data by iteratively grouping samples according to their similarity using typically an Euclidean distance. The relevance of this family of algorithms has been reported in a variety of remote sensing applications, including delimitation of climatic regions [Rhee et al., 2008], identification of snow cover [Farmer et al., 2010], characterization of forest habitats [Bunting et al., 2010] or definition of oceanic ecosystems [Hardman-Mountford et al., 2008]. For specific tasks of image segmentation, spatial coherence of the segments is crucial. Therefore, contextual constraints have been included into the base algorithms. As an example, multistage restricted methods are proposed in [Lee, 2004, Lee and Crawford, 2004] to perform first a region growing segmentation to ensure spatial contiguity of the segments, and then to perform classical linkage to merge the most similar segments. In [Marcal and Castro, 2005], several informative features are included in the aggregation rule: spatial criteria accounting for cluster compactness, cluster size, and the part of their boundaries that two clusters share ensure spatially coherent clusters. Fractal dimension is used as a criterion of homogeneity in [Baatz and Schäpe, 2000]. Other ways to ensure spatial coherence is to use region-growing algorithms: Tarabalka et al. [2010] utilize random forests for that purpose. Watershed segmentation on the different resolutions is proposed in [Hasanzadeh and Kasaei, 2010], where it is used in conjunction with fuzzy partitioning to account for connectedness of segments.

4.2.3 A SUPERVISED CLASSIFICATION EXAMPLE

In the following we provide a comparison of five supervised methods for land-use mapping in an urban environment. The compared methods are LDA, a standard classification tree, k-nearest neighbors, SVM, and the multilayer perceptron. Urban areas are the most challenging problems in very high resolution image classification because they present a great diversity and complexity in the objects represented [Licciardi et al., 2009b]. In particular, the geometry and context that can be retrieved is essential to account for different land uses that are composed by the same material (thus the same spectral behavior at the pixels level) but vary as a function of its context.

To stress this point, Tables 4.1 and 4.2 show the numerical results obtained by the methods with and without contextual features on the QuickBird image of Brutisellen (Switzerland) described in Section 4.1.2. For the nine classes of interest, small differences can be made between classes 'Roads' and 'Parkings' (both are asphalt) and between 'Trees' and 'Meadows'. To enhance the performance in these classes, morphological top-hat features (see Chapter 3) have been computed for the four bands and stacked to the multispectral bands before training the classifiers.

The numerical results show a few clear trends: the excellent performances of nonparametric methods such as support vector machines and neural networks, the good performance of lazy learners as k-nearest neighbors when a sufficient number of labeled examples is available, and the poor performance of linear parametric classifiers as the LDA. The addition of contextual information is beneficial for all models, whose accuracy increases between 5% and 10%. Interestingly, note that without contextual information, SVM and NN are statistically inferior to k-NN, because they attempt to generalize classes that cannot be separated with spectral information only. On the other hand, k-NN outperforms the other methods when using non-discriminative information, since the model does not rely on a trained model and analyzes each pixel by its direct neighbors.

The classification maps obtained with the different approaches with contextual information are illustrated in Fig. 4.3: among all classifiers presented, SVM and k-NN return the closest maps to the expected result, and can efficiently detect all major structures of the image. This demonstrates the suitability of nonparametric models for supervised image classification. For this realization, the MacNemar's test confirmed these visual estimation of the quality: SVM map is significantly better than the others, followed by the k-NN and NN maps.

Table 4.1: Comparison of supervised classification algorithms. Mean overall accuracy and Kappa statistics (see Section 4.1.3), along with standard deviation of five independent experiments. In bold, the best performing method. Underlined, results with average Kappa statistically inferior to the best method.

Training pixels		OA [%]					Kappa				
		LDA	Trees	k-NN	SVM	MLP	LDA	Trees	k-NN	SVM	MLP
115	μ	60.43	68.62	68.43	**74.99**	72.94	0.53	0.61	0.61	**0.69**	0.67
	σ	(5.13)	(3.85)	(1.63)	(2.25)	(1.55)	(0.06)	(0.05)	(0.02)	(0.03)	(0.02)
255	μ	60.19	71.25	73.65	**77.31**	76.32	0.53	0.64	0.67	**0.72**	0.71
	σ	(3.25)	(1.79)	(3.79)	(1.23)	(1.20)	(0.03)	(0.02)	(0.05)	(0.02)	(0.02)
1155	μ	62.82	76.78	**80.92**	79.49	79.41	0.56	0.71	**0.76**	0.74	0.74
	σ	(2.08)	(0.90)	(0.47)	(0.73)	(0.38)	(0.02)	(0.01)	(0.01)	(0.01)	(0.01)
2568	μ	62.68	78.59	**81.38**	80.42	79.42	0.56	0.74	**0.77**	0.76	0.74
	σ	(1.94)	(0.32)	(0.24)	(0.34)	(1.09)	(0.02)	(0.01)	(0.01)	(0.01)	(0.01)

Table 4.2: Comparison of supervised classification algorithms with contextual information. Mean overall accuracy and Kappa statistics (see Section 4.1.3), along with standard deviation of five independent experiments. In bold, the best performing method. Underlined, results with average Kappa statistically inferior to the best method.

Training pixels		OA [%]					Kappa				
		LDA	Trees	k-NN	SVM	MLP	LDA	Trees	k-NN	SVM	MLP
115	μ	72.93	71.00	75.69	**83.37**	77.37	0.67	0.65	0.70	**0.80**	0.72
	σ	(2.85)	(2.97)	(1.28)	(2.40)	(2.48)	(0.03)	(0.03)	(0.02)	(0.03)	(0.03)
255	μ	77.23	73.47	80.53	**85.91**	80.61	0.72	0.68	0.76	**0.83**	0.76
	σ	(1.41)	(1.64)	(1.34)	(1.94)	(0.99)	(0.02)	(0.02)	(0.02)	(0.02)	(0.01)
1155	μ	78.35	80.45	87.32	**88.03**	84.29	0.74	0.76	0.84	**0.85**	0.81
	σ	(0.69)	(0.73)	(0.63)	(1.68)	(1.77)	(0.01)	(0.01)	(0.01)	(0.02)	(0.02)
2568	μ	78.61	81.59	**87.26**	87.17	85.10	0.74	0.77	**0.84**	**0.84**	0.82
	σ	(0.57)	(0.89)	(0.61)	(0.85)	(1.05)	(0.01)	(0.01)	(0.01)	(0.01)	(0.01)

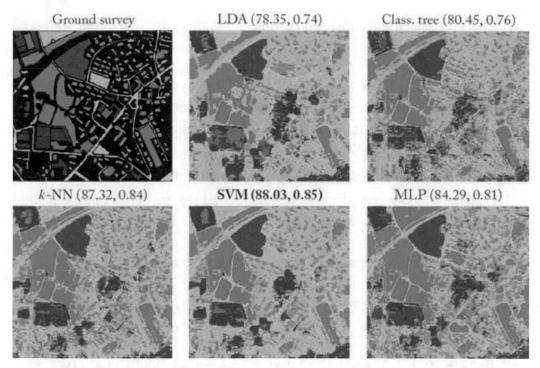

Figure 4.3: Classification maps of the experiments using contextual information and 1155 training pixels. Best results are shown in parentheses in the form of (OA [%], κ).

4.3 CHANGE DETECTION

The second family of classification problems discussed in this chapter is related to the detection of class transitions between a pair of co-registered images, also known as *change detection* [Radke et al., 2005, Singh, 1989]. Change detection is attracting an increasing interest from the application domains, since it automatizes traditionally manual tasks in disaster management or developments plans for urban monitoring. We should also note that multitemporal classification and change detection are very active fields nowadays because of the increasing availability of complete time series of images and the interest in monitoring Earth's changes at local and global scales. In a few years from now, complete constellations of civil and military satellites sensors will deliver a much higher number of revisit time images. To name a few, the ESA's Sentinels[3] or NASA's A-train[4] programs are expected to produce near real-time scientific coverage of the globe in the next decades.

Three types of products are commonly studied in the change detection framework [Coppin et al., 2004, Singh, 1989]: binary maps, detection of types of changes and full multi-class change maps, thus including classes of changes and unchanged land-cover classes. Each type of product can be achieved using different sources of information retrieved from the initial spectral images at time instants t_1 and t_2. Therefore, the input vector $\mathbf{x}$ can be either composed by both separate datasets $\mathbf{x}_{(1)}$ and $\mathbf{x}_{(2)}$, or by a combination of the features of two images taken at time instants t_1 and t_2. Singh [1989] underlines that changes are likely to be detected if their changes in radiance are larger than changes due to other factors such as differences in atmospheric conditions or sun angle. Many data transformations have been used to enhance the detection of changes:

- the stacking of feature vectors $\mathbf{x}_{\text{stack}} = [\mathbf{x}_{(1)}, \mathbf{x}_{(2)}]$, mainly used in supervised methods,

- the difference image $\mathbf{x}_{\text{diff}} = |\mathbf{x}_{(1)} - \mathbf{x}_{(2)}|$, where unchanged areas show values close to zero,

- image ratioing $\mathbf{x}_{\text{ratio}} = \frac{\mathbf{x}_{(1)}}{\mathbf{x}_{(2)}}$, where unchanged areas show values close to one,

- data transformations as principal components (see Chapter 3), where changes are grouped in the components related to highest variance,

- physically based indices as NDVI (see Chapters 3 and 6), useful to detect changes in vegetation.

However, these transforms do not cope with problems of registration noise or intraclass variance. Therefore, algorithms exploiting contextual filtering (see Chapter 3) have been also considered in the literature. In the following section, a brief review of different strategies is reported.

4.3.1 UNSUPERVISED CHANGE DETECTION

Unsupervised change detection has been widely studied, mainly because it meets the requirements of most applications: i) the speed in retrieving the change map and ii) the absence of labeled information. However, the lack of labeled information makes the problem of detection more difficult and thus unsupervised methods typically consider binary change detection problems. To do so, the multitemporal data are very often transformed from the original to a more discriminative domain using feature extraction methods, such as PCA or wavelets, or converting the difference image to polar coordinates, i.e., magnitude and angle. In the transformed space, radiometric changes are assessed in a pixel-wise manner.

The most successful unsupervised change detector until present remains the change vector analysis (CVA, [Bovolo and Bruzzone, 2007, Malila, 1980]): a difference image is transformed to polar coordinates, where a threshold discriminates changed from unchanged pixels. In [Dalla Mura et al., 2008], morphological operators were successfully applied to increase the discriminative power of the CVA method. In [Bovolo, 2009], a contextual parcel-based multiscale approach to unsupervised change detection was presented. Traditional CVA relies on the experience of the researcher for the threshold definition, and is still on-going research [Chen et al., 2010a, Im et al., 2008]. The method has been also studied in terms of sensitivity to differences in registration and other radiometric factors [Bovolo et al., 2009].

Another interesting approach based on spectral transforms is the multivariate alteration detection (MAD, [Nielsen, 2006, Nielsen et al., 1998]), where canonical correlation is computed for the points at each time instant and then subtracted. The method consequently reveals changes invariant to linear transformations between the time instants. Radiometric normalization issues for MAD has been recently considered in [Canty and Nielsen, 2008].

Clustering has been used in recent binary change detection. In [Celik, 2009], rather than converting the difference image in the polar domain, local PCAs are used in subblocks of the image, followed by a binary k-means clustering to detect changed/unchanged areas locally. Kernel clustering has been studied in [Volpi et al., 2010], where kernel k-means with parameters optimized using an unsupervised ANOVA-like cost function is used to separate the two clusters in a fully unsupervised way.

Unsupervised neural networks have been considered for binary change detection: Pacifici et al. [2009b] proposed a change detection algorithm based on pulse coupled neural networks, where correlation between firing patterns of the neurons are used to detect changes. In [Pacifici et al., 2010], the system is successfully applied to earthquake data. In Ghosh et al. [2007], a Hopfield neural network, where each neuron is connected to a single pixel is used to enforce neighborhood relationships.

4.3.2 SUPERVISED CHANGE DETECTION

Supervised change detection has begun to be considered as a valuable alternative to unsupervised methods since the advent of very high resolution (VHR) sensors made obtaining reliable ground truth maps by manual photointerpretation possible. Basically, two approaches are found in the literature: the post-classification comparison (PCC, [Coppin et al., 2004]) and the multitemporal classification [Singh, 1989], which can be roughly divided into direct multi-date classification and difference analysis classification. The next paragraphs review these two approaches.

PCC methods are based on individual classifications of the datasets at each time instant. Once the classification maps are obtained, the results are computed by subtracting them. Traditional PCC is often inefficient because it suffers from the cumulation of the errors made by individual classifications. This ends up in salt-and-pepper change detection maps, where several changes in unchanged areas (typically at the borders between classes) are observed. Purple areas in Fig. 4.4(a) are areas where these errors occur. To cope with these errors, postclassification masking has been proposed in [Pacifici et al., 2007]: a neural network trained to solve a binary change vs. no-change problem masks the unchanged areas, thus decreasing the post-classification errors. In [Chen et al., 2010a], CVA is used for this same purpose and also for discriminating the type of changes. Although very efficient, these approaches consider consequences rather than causes of the PCC problems.

Multitemporal classification has been considered in three settings: by using stacked vectors x_{stack} before training the classifier (DMC), by computing a difference image x_{diff} before analysis (DIA), or by designing classifier-specific measures to cope for both time instants simultaneously. The main difference between the two first is that DMC can return a full land-cover change maps, where both the changed and unchanged land-cover types are described, while DIA cannot differentiate among unchanged land-cover types, as they all reduce to spectra with nearly null components. Regarding the third setting, Camps-Valls et al. [2008a] studied the effect of summation, difference and ratio kernels to solve change detection problems with binary, multiclass and one-class SVM. In this work, the use of composite kernels also allows one to combine information from different time instants and multisource images (e.g., optical and radar images). Actually, we should note that the use of spatial and contextual features is currently poorly documented in supervised change detection. Only few studies can be found (see, for instance, Pagot and Pesaresi [2008], Volpi et al. [2011a]). However, they all confirm the interest of filtering the magnitude of the difference image to reduce radiometric differences and registration errors.

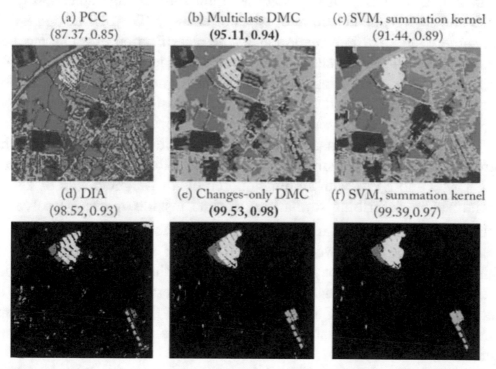

(a) PCC (87.37, 0.85)   (b) Multiclass DMC (95.11, 0.94)   (c) SVM, summation kernel (91.44, 0.89)

(d) DIA (98.52, 0.93)   (e) Changes-only DMC (99.53, 0.98)   (f) SVM, summation kernel (99.39, 0.97)

Figure 4.4: Supervised change detection results. (a) PCC (purple areas are misclassified areas, where unobserved transitions are detected); (b) DMC stacked multiclass; (c) SVM with summation kernel (SVM specific); (d) DIA; (e) DMC stacked without differences in the unchanged class; (f) SVM with summation kernel without differences in unchanged areas. Best results are shown in parentheses in the form of (OA [%], κ).

Finally, the scarceness of labeled examples has been tackled with target detection models [Camps-Valls et al., 2008a] and semisupervised learning [Bovolo et al., 2008, Gómez-Chova et al., 2010, Muñoz-Marí et al., 2010]. These two aspects will be detailed in Sections 4.5.1 and 4.4, respectively.

4.3.3 A MULTICLASS CHANGE DETECTION EXAMPLE

In this section, four approaches for supervised change detection are compared for the VHR image of Brutisellen, described in Section 4.1.2. In particular, we used PCC, DMC, SVM with summation kernel [Camps-Valls et al., 2008a] and DIA in two settings: multiclass change detection (PCC, DMC and SVM) and detection of types of changes (DIA, DMC and SVM). The same contextual information as in the classification example has been used, and four classes of change have been considered, as illustrated in Fig. 4.2(f): meadows to buildings, soil to buildings, meadows to roads and soil to meadows. The change detection results are illustrated in Fig. 4.4.

The change detection maps of Fig 4.4 confirm the main disadvantage of PCC: many false alarms are detected (in purple in Fig 4.4(a)) since they correspond to single classification errors and differences in object borders. Both DMC and SVM with summation kernels (a weighted sum of kernels specialized in describing similarities at time instants t_1 and t_2) solve this problem and return land-use maps accounting for the changes that have occurred between the two dates. DMC seems more dependent on the registration noise (see, for instance, the path in the middle of the meadow on the left side of the image). The summation kernel can better exploit the specific spatial context of the pixels and reduces the misclassification rate, mainly due to the fact that estimates similarities separately for each time instant.

Considering the detection of types of changes, the three algorithms detect the main patches of change in the image. However, DIA returns a noisy result since it is based on a difference image and nonzero vectors may be erroneously interpreted as change. On the contrary, DMC and even more the summation kernel return improved description of the unchanged class, while losing precision in the description of the changed classes. According to the analyst's interest (detecting clearly changed areas or fine detection of type of changed objects), DIA or SVM with specialized kernels should be the preferred methods in this problem.

4.4 DETECTION OF ANOMALIES AND TARGETS

This section is devoted to the field of automatic target detection. Some applications relying on remote sensing data do not need the definition of many classes of interest. Actually, very often, they only need to discriminate a single class from the rest, i.e., the background [Ahlberg and Renhorn, 2004]. Sometimes, there is no knowledge about the signature of the class of interest, but it is known that this signature is different from the rest. In this case, a model is built by looking for signatures that deviate from a model of the background. Large deviations from the mean are called *anomalies*, and the models based on this principle are referred to as *anomaly detectors*. Anomaly detectors can be very difficult to calibrate, especially when there is a strong spectral variability in the image. This field is related to the identification of endmembers that will be treated in Chapter 5.

Most of the time we are aware of the type of class we are looking for. In this case, we may have some labeled pixels of the class of interest or we can rely on databases (often called *spectral libraries*) of class-specific spectral signatures. In this case, the problem reduces to defining the background class and then detecting spectral signatures that are closer to the known signatures than to the background. Algorithms searching for these signatures, or *targets*, are referred to as *target detectors*. Target detection is of particular interest when the targets are subpixel targets or buried objects, thus known but difficultly detectable on the image [Chang, 2003, Chang and Heinz, 2000]. Subpixel target detection, or *unmixing* will be detailed in Chapter 5. In the rest of this chapter, we will consider anomalies and targets bigger than a pixel. Table 4.3 summarizes the anomaly and target detection methods presented in the next sections.

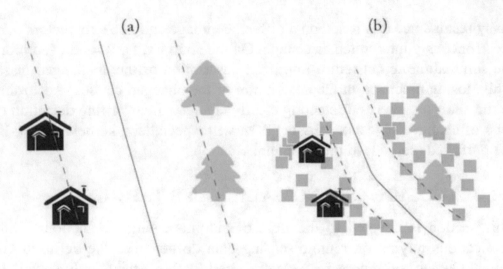

Figure 4.5: Semisupervised learning: (a) the model obtained with the labeled samples only is not representative of the global probability density function (PDF). (b) By integrating the PDF statistics, the model adapts to the unlabeled data.

Table 4.3: Taxonomy of anomaly and target detectors (adapted from Ahlberg and Renhorn [2004] and Kwon and Nasrabadi [2007a]).

	Background model		
	Gaussian	Subspace	Nonparametric
Anomaly detectors	RX, GMM	DFFS	One Class SVM, SVDD
Target detectors	SAM, AMF	ASD, OSP, SMF, MSD	One Class SVM, SVDD, MIL, KASD, KOSP, KSMF, KASD

4.4.1 ANOMALY DETECTION

When the signature of the class to be detected is unknown, anomaly detection algorithms are used. Anomaly detection reduces to assuming (and/or defining) a model for the background class, and then looks for the test samples lying far from the mean of the background distribution.

Depending on the assumptions made on the background, different anomaly detectors can be used. If the background is modeled as a Gaussian distribution, the Reed-Xiaoli (RX) [Lu et al., 1997, Reed and Yu, 1990] detector can be used. The RX detector detects outliers by computing a

Mahalanobis distance between the considered pixel and the background mean vector. If the background is believed to be more complex, for instance multi-Gaussian, it can be modeled with a Gaussian mixture model [Beaven et al., 2000]. Application of these principles to the detection of small objects in series of remote sensing images can be found in [Carlotto, 2005]. More recently, the RX detector has been extended to its nonlinear kernelized version in [Kwon and Nasrabadi, 2005].

Another possibility is to model the background as a linear subspace described by a set of linearly independent basis vectors. Assuming such a model, any vector in the subspace can be written as a linear combination of these basis functions. As seen before, these basis functions are known in remote sensing as *endmembers*. The Distance From Feature Space (DFFS) detector computes the distance of the pixel from the background as the squared norm of the residual among the pixel and its projection in the linear subspace. Kernel PCA has been used in [Gu et al., 2008] to project data into a nonlinearly independent space: the RX algorithm is then successfully applied on the component maximizing a measure of skewness and kurtosis.

If no global assumptions are made and one assumes that only local samples are important, nearest neighbor detectors can be used instead. In this case, the distance of the pixels to their nearest neighbors in the background are computed: the weighted mean distance is then thresholded to decide whether the pixel is an outlier or not. Finally, nonparametric models can be used to model the background, as in [Banerjee et al., 2009], where a support vector domain description (SVDD) is used to model the background and then detect outliers, which are the pixels lying outside the region of support of the background pixels. We will use this model in Section 5.4.

Another recent field of investigation in this direction is *anomalous change detection* [Theiler, 2008b]: in this field, one looks for changes that are interestingly anomalous in multitemporal series, and tries to highlight them in contrast to acquisition condition changes, registration noise, or seasonal variation [Theiler et al., 2011]. Methods span from chronochrome maps [Schaum and Stocker, 1997a,b], where the best linear estimator for the background in the second time instant is found, to covariance equalization methods [Schaum and Stocker, 1994], methods estimating the mutual information between the appearance of given values in the two acquisitions [Theiler and Perkins, 2006] and pervasive methods [Theiler, 2008a], where noise is added to an image to account for misregistration.

4.4.2 TARGET DETECTION

When the target is partially or entirely known, this additional knowledge can be used to design more accurate target detectors: rather than assessing the distance from a background distribution, the pixel is compared against target and background and then assigned to the closest class.

When the background and target are assumed to be Gaussian-distributed, a likelihood function can be used to decide whether a pixel belongs to one or the other. Such likelihood maximizes the distance to the background and minimizes the distance to the target simultaneously. Using a Mahalanobis distance, this target detector is called the adaptive matched filter (AMD). Another possibility is to use the spectral angle mapper (SAM) as a measure of distance. Such measure assesses the angle between the target and the pixel to be evaluated with a dot product.

When using a linear subspace for the background model, the orthogonal space projection (OSP) detector is one of the most famous models used in remote sensing [Lu et al., 1997, Reed and Yu, 1990]: in this case, the subspace of the background basis functions is removed from the analyzed pixel, thus leaving only the part related to the target signature. This reduced spectral signature is then matched with the target signature: if the match exceeds a given threshold, the pixel is considered as target. OSP detectors have been extended to the nonlinear case using kernel methods in [Kwon and Nasrabadi, 2005]: the kernel orthogonal space projection (KOSP) calculates the OSP detector in a high dimensional feature space and can handle non-Gaussian complex background models. In [Capobianco et al., 2009], a semisupervised version of KOSP is proposed using graphs to preserve the structure of the manifold (the field of semisupervised learning was treated in Section 4.5.1). Other popular linear target detection approaches include the matched subspace detector (MSD) [Scharf and Friedlander, 1994], the spectral matched filter (SMF) [Robey et al., 1992], and the adaptive subspace detectors (ASD) [Kraut et al., 2001]: Kwon and Nasrabadi [2007a] present a review on these target detectors and their kernel counterparts: see kernel matched subspace detector (KMSD Kwon and Nasrabadi [2006b]), kernel spectral matched filter (KSMF Kwon and Nasrabadi [2007b]) and kernel adaptive subspace detector (KASD Kwon and Nasrabadi [2006a]). Other methods that project data into an orthogonal low-dimensional space can be used to separate target from background: principal components analysis, Fisher's discriminant and eigenspace separation, along with their kernel counterparts can be used for this purpose [Nasrabadi, 2009].

Another possibility is to reduce the target detection problem to a single class classification problem. Nonparametric methods, such as SVDD [Muñoz-Marí et al., 2007] and the one-class SVM [Camps-Valls et al., 2008a, Muñoz-Marí et al., 2010, Wang et al., 2006] can also be used for target detection: in this case, the known target pixels are used to train the model. A traditional application for this type of approach is the detection of landmines: among others, contextual-aware approaches such as Hidden Markov Models [Gader et al., 2001, Zhu and Collins, 2005], 3-D histograms [Frigui and Gader, 2009], random sets [Bolton and Gader, 2009] and multiple instance learning [Bolton and Gader, 2010] have been proposed for this purpose.

4.4.3 A TARGET DETECTION EXAMPLE

In the following, four target detectors are compared in the task of detecting the class 'Soil' of the Zürich image. In Figs. 4.6(a)-(d), the detections of OSP and KOSP with varying parameters are illustrated. OSP returns a decision function strongly contaminated by noise, and it is difficult to apply a threshold to efficiently isolate the target. On the contrary, a well-calibrated KOSP detector can result into correct detection, as shown in Fig. 4.6(c). However, KOSP appears as being strongly influenced by the choice of the kernel parameter: a too small σ results in overfitting the target endmember signature (Fig. 4.6(b)), while a too large kernel bandwidth σ returns a completely wrong detection (Fig. 4.6(d)).

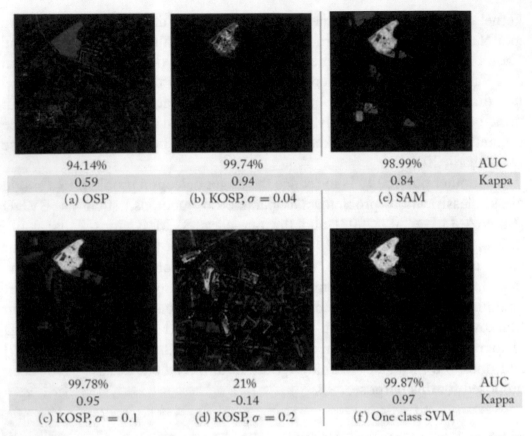

94.14%	99.74%	98.99%	AUC
0.59	0.94	0.84	Kappa
(a) OSP	(b) KOSP, $\sigma = 0.04$	(e) SAM	

99.78%	21%	99.87%	AUC
0.95	-0.14	0.97	Kappa
(c) KOSP, $\sigma = 0.1$	(d) KOSP, $\sigma = 0.2$	(f) One class SVM	

Figure 4.6: Comparison of algorithms for target detection of the class 'Soil' (in brown): values of the discriminative function for (a) OSP, (b-d) KOSP with different kernel parameters, (e) SAM detector and (f) One class SVM using 50 pixels representing the target. Kappa is maximized by varying the detection threshold.

The SAM detector returns a good reconstruction of the target, but with strong false alarms in the bottom left side of the image, where the roof of the commercial center saturates the sensor and thus returns a flat spectrum for its pixels in the morphological features. As a consequence, both the target and the roof vectors have flat spectra and their spectral angle is almost null. The one-class SVM returns a highly discriminative function, where the target and background are nicely discriminated. The rate of false alarms is thus minimized, returning the best numerical and visual result among the compared methods.

4.5 NEW CHALLENGES

This section deals with emerging problems in remote sensing image classification. Now that the methodologies have reached a great degree of sophistication and precision, new problems related to image acquisitions, atmospheric conditions and representativity of the training samples are being tackled in the remote sensing image processing community. This section briefly resumes some of the most exciting fields of research nowadays.

4.5.1 SEMISUPERVISED LEARNING

In supervised remote sensing image classification problems, the information conveyed by the unlabeled pixels is disregarded. However, a large amount of unlabeled pixels is available at no additional cost. Recently, unlabeled data have been used to increase robustness of classifiers. Semisupervised methods fuse the advantages of supervised and unsupervised methods: while the first gives the true class membership of a few pixels, the other models the global structure of the image and enforces smoothness in the decision function (Figure 4.5). As a consequence, classification maps overfit more difficultly in cases of small training sets.

The first successful attempts to semisupervised learning are found in [Tadjudin and Landgrebe, 2000], where unlabeled examples are used to update the Maximum Likelihood classifier's parameters using the expectation maximization algorithm. In [Jackson and Landgrebe, 2001], this strategy is made more stable by adding decreased weights to the unlabeled examples. However, this solution only applies when a parametric model can be established for the data (Gaussian mixtures, for the cases presented above). These *generative* approaches have been abandoned due to the rigid assumptions made. Actually, more recent research has focused on nonparametric classifiers and in particular on kernel and graph-based methods. The effect of regularizers accounting for misclassification of semilabeled examples are studied in [Dundar and Langrebe, 2004] for kernel Fisher discriminant classification and in [Bruzzone et al., 2006, Chi and Bruzzone, 2007] for SVM. In both cases, the cost of semilabeled examples being on the wrong side of the classification hyperplane is penalized, thus regularizing the approximated solution obtained with the labeled examples only. Graphs are used efficiently to add manifold regularizers to the SVM classifier in [Camps-Valls et al., 2007, Gómez-Chova et al., 2008] for hyperspectral classification and in [Capobianco et al., 2009, Muñoz-Marí et al., 2010] for target detection. Tuia and Camps-Valls [2009] propose to deform the training SVM kernel with a likelihood kernel computed using a clustering algorithm; the deformed kernel is then used with the standard SVM solver. Finally, some efforts have been made towards adapting neural networks to semisupervised learning, which have reported good results in large scale scenarios [Ratle et al., 2010].

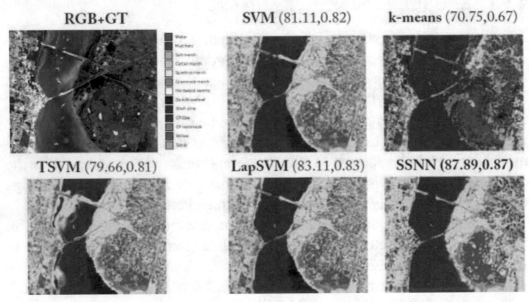

Figure 4.7: RGB composition along with the ground truth (GT) available, and the classification maps with the best SVM, k-means, TSVM, LapSVM, and SSNN for the KSC image ($l = 200$, $u = 1000$). Overall accuracy and kappa statistic are given in brackets.

4.5.2 A SEMISUPERVISED LEARNING EXAMPLE

Figure 4.7 shows an illustrative example of semisupervised classification. We show accuracy values and classification maps for standard unsupervised k-means, supervised SVM, and semisupervised algorithms, such as the transductive SVM (TSVM), the Laplacian SVM (LapSVM), and the semisupervised neural network (SSNN). The LapSVM improves the results of the TSVM, while more homogeneous areas and better classification maps are observed for the SSNN. See, for instance, the high spatial coherence on the Indian River Lagoon (IRL) waters, in which TSVM/LapSVM yield a noisy classification, while SSNN achieves spatially homogeneous results. This suggests that exploiting unlabeled samples properly is crucial for modeling the data complexity (e.g., similar water spectra are confused when using a low number of unlabeled samples). Another interesting example is that of the inhomogeneous area of impounded estuarine wetlands at south-east of the image. Here, small marsh class are correctly detected with the SSNN method, while over-regularized solutions are obtained with TSVM/LapSVM probably due to the weak support used for data modeling.

4.5.3 ACTIVE LEARNING

An interesting alternative to semisupervised learning to correct a suboptimal model is to consider interaction with the user. *Active learning* defines a query function, or *heuristic*, that ranks the unlabeled examples by their uncertainty and then samples and labels the pixels showing maximal uncertainty. By doing so iteratively, the model is forced to resolve the high uncertainty areas near the classification boundary (Figure 4.8), while keeping the good performance on the already known areas. The interest of active learning algorithms lies in the definition of the heuristic: the more the heuristic is capable to find informative uncertain pixels, the faster the model will improve its performance. The aim is to minimize the number of queries necessary to reach a desired classification result.

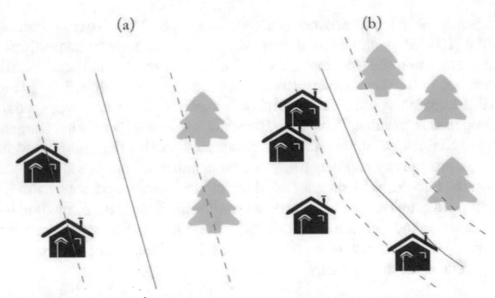

Figure 4.8: Active learning: (a) the model obtained with the labeled samples only is inaccurate in uncertain areas near the classification boundary. (b) By labeling new samples close to the classification boundary, the model is improved in complex areas.

The uncertainty of an unlabeled pixel can be measured in many ways: for instance, the predictions of a committee of learners using bootstrapped training sets (EBQ, [Tuia et al., 2009b]) or subsets of the feature space (AMD, [Di and Crawford, 2011]) can be used. The more the predictions of the committee disagree, the more uncertain the committee is about the label of this pixel. Otherwise, posterior probabilities of the class assignment for the pixel candidate $p(y | \mathbf{x})$ can be used: heuristics using this principle can consider single sample posteriors directly (BT, [Luo et al., 2005]) or evaluating divergence between distributions with and without the candidate (KL-max, [Rajan et al., 2008]).

However, most of active learning heuristics proposed in the literature are specific to SVM classifiers (see Table 4.4 for a summary of active learning algorithms). SVM are naturally well-adapted to active learning, since the decision function of the model, which can be interpreted as the distance of an unlabeled pixel to the separating hyperplane, can be used to assess the uncertainty of the class assignment of this pixel. The further away from the boundary, the more certain the class assignment will be. Heuristics exploiting this fact are the margin sampling (MS, [Mitra et al.,

2004]), where pixels minimizing this function are selected and the multiclass level uncertainty (MCLU, [Demir et al., 2011]), where the difference between the two most certain classes is used instead. In [Pasolli et al., 2011], the heuristic proposed is specialized at looking for potential support vectors, by adding a second binary classifier discriminating potential support vectors.

Another issue active learning algorithms deal with is the selection of several samples – or batches – at each iteration. Once capable of selecting the most uncertain pixels, simple strategies as MS, EQB or MCLU risk to sample pixels that are informative with respect to the current model, but also redundant between each other. To avoid this, second stage heuristics constraining to have diverse samples in the batch have been proposed, mainly for SVM. To this end, diversity has been considered between candidate pixels [Demir et al., 2011, Ferecatu and Boujemaa, 2007] or with respect to the current model [Tuia et al., 2009b] or both [Volpi et al., 2011b]. In these works, mixed heuristics with an uncertainty and a diversity part are used to have the most informative batch of samples.

Recent papers deal with new applications of active learning algorithms: in [Li et al., 2010, Liu et al., 2008b], active learning is used to select the most useful unlabeled pixels to train a semisupervised classifier, while in [Jun and Ghosh, 2008, Tuia et al., 2011a] active queries are used to correct for dataset shift in different areas of images (see Section 4.5.5). Finally, active learning has been used to prune hierarchical clustering for semisupervised image segmentation [Tuia et al., 2010b]: in this case, no supervised model is iteratively optimized, rather the cut in the hierarchical clustering tree, by optimizing a cost function depending on the confidence of the cluster label assignment.

The great challenge of active learning for future implementation is to include spatial constraints in the heuristics, in order to allow the model to plan the planning campaign including the terrain constraints influencing the cost of discovering every label. A first application in this direction can be found in [Liu et al., 2008a].

4.5.4 AN ACTIVE LEARNING EXAMPLE

In Fig. 4.9, four active learning heuristics are compared in terms of the number of label queries: The Entropy Query-by-Bagging (EQB, Tuia et al. [2009b]), the Margin Sampling (MS, Mitra et al. [2004]), the Multiclass-level uncertainty with angle based diversity (MCLU-ABD Demir et al. [2011]), and the breaking ties (BT, Luo et al. [2005]). From the learning curves, it is clear that active selection of pixels leads to faster learning rates than random selection (black dashed line in the figure), with gains between 2% and 3% in accuracy. With actively selected training sets of 500-600 pixels, the model already performs as the best result reported in Table 4.2 (page 66), where 2568 samples were randomly selected.

4.5.5 DOMAIN ADAPTATION

As stressed in Chapter 1, remote sensing data show a strong temporal component. The spectra acquired by the sensor account simultaneously for the surface reflectance and for the illuminant impacting the surface at that precise moment. Both sources are affected by the temporal component: seasonal changes deform the acquired spectral information of vegetation surfaces, for instance according to plants growth cycle, phenology, or crop periods. Regarding the illuminant, images taken at different time instants show different illumination conditions, different atmospheric conditions, and differences in the shadowing patterns. These issues lead to *shifts* in the data distribution, as depicted in Fig. 4.10.

Table 4.4: Summary of active learning algorithms Tuia et al. [2011c] (B: binary, M: multiclass, c: number of candidates, p: members of the committee of learners.

Heuristic	Reference	Batches	Type	Classifier	Diversity S	Diversity SVs	Models to train
EQB	Tuia et al. [2009b]	✓	M	*All*	×	×	p models
AMD	Di and Crawford [2011]	✓	M	*All*	×	×	p models
MS	Mitra et al. [2004]	✓	B	SVM	×	×	Single SVM
MCLU	Demir et al. [2011]	✓	M	SVM	×	×	Single SVM
SSC	Pasolli et al. [2011]	✓	B	SVM	×	×	2 SVMs
cSV	Tuia et al. [2009b]	✓	B	SVM	≈	✓	Single SVM + distances to support vectors
MOA	Ferecatu and Boujer [2007]	✓	B	SVM	✓	×	Single SVM + distances to already selected pixels
MCLU-ABD	Demir et al. [2011]	✓	M	SVM	✓	×	Single SVM + distances to already selected pixels
MCLU-ECBD	Demir et al. [2011]	✓	M	SVM	✓	×	Single SVM + nonlinear clustering of candidates
hMCS-i	Volpi et al. [2011b]	✓	M	SVM	✓	✓	Single SVM + nonlinear clustering of candidates and SVs
KL-max	Rajan et al. [2008]	×	M	$p(y\|\mathbf{x})$	×	×	$(c-1)$ models
BT	Luo et al. [2005]	✓	M	$p(y\|\mathbf{x})$	×	×	Single model

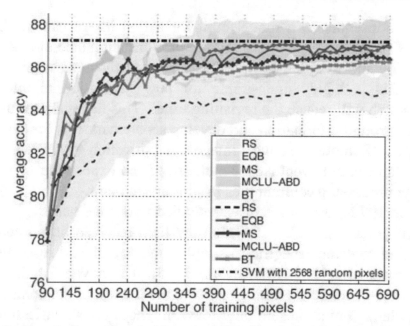

Figure 4.9: Comparison of active learning algorithms. In black the lower (random selection, RS) and upper (SVM with 2568 randomly selected training pixels) bounds.

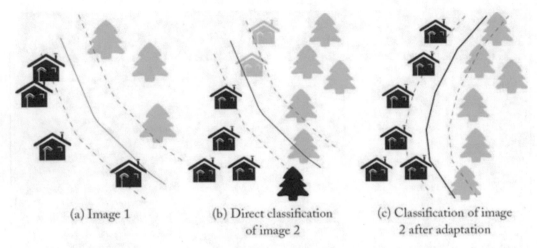

| (a) Image 1 | (b) Direct classification of image 2 | (c) Classification of image 2 after adaptation |

Figure 4.10: The problem of adaptation: in (a) a model is optimized for the first image to recognize two classes. When applied to the second image directly (b) several pixels are misclassified, since the distribution has shifted between the two acquisitions (in this case, the shift is represented by a rotation). Adapting the model (c) can solve the problem correctly.

As a consequence, a model built for a first image can be difficultly applied to an image taken at the second time instant, since the reflectances described by the first one do not correspond to those represented by the second. This is highly undesirable for new generation remote sensing images, which will reduce the revisiting times. This fact makes it unfeasible to provide ground information and develop separate models for every single acquisition. Remote sensing image classification requires *adaptive* solutions, and could eventually benefit from the experience of the signal and image processing communities in similar problems.

In the 1970s, this problem was studied under the name of "signature extension" [Fleming et al., 1975], but the field received little interest until recently and remained specific to simple models and mid resolution land-use applications, as in [Olthof et al., 2005]. The advent of the machine learning field of transfer learning renewed the interest of the community for these problems, but at present only few references for classification can be found. The pioneer work considering unsupervised transfer of information among domains can be found in [Bruzzone and Fernández-Prieto, 2001], where the samples in the new domain are used to re-estimate the classifier based on a Gaussian mixture model. This way, the Gaussian clusters optimized for the first domain are expected to match the data observed in the second domain. In [Rajan et al., 2006b] randomly generated classifiers are applied in the destination domain. A measure of the diversity of the predictions is used to prune the classifiers ensemble. In [Bruzzone and Marconcini, 2009], a SVM classifier is modified by adding and removing support vectors from both domains in an iterative way: the model discards contradictory old training samples and uses the distribution of the new image to adapt the model to the new conditions. In [Bruzzone and Persello, 2009], spatial invariance of features into a single image is studied to increase the robustness of a classifier in a semisupervised way. In Gómez-Chova et al. [2010], domain adaptation is applied to a series of images for cloud detection by means of matching of the first order statistics of the data clusters in source and destination domains is performed in a kernel space: two kernels accounting for similarities among samples and clusters are combined and the SVM weights are computed using the labels from samples in the first domain. Finally, Leiva-Murillo et al. [2010] include several tasks in a SVM classifier, one for each image, showing the interest of simultaneously considering all the images to achieve optimal classification.

If it is possible to acquire a few labeled pixels on the second image, active learning schemes (see Section 4.5.3) can be used to maximize the ratio of informativeness per acquired pixels: approaches in this sense can be found in [Rajan et al., 2006a], where active learning is used to find shifts in uncertainty areas and in [Tuia et al., 2011a], where a feature space density-based rule is proposed to discover new classes that were not present in the first acquisition. Finally, Persello and Bruzzone [2011] propose a query function removing labeled pixels of the original domain that do not fit the adapted classifier.

4.6 SUMMARY

One of the main end-user applications of remote sensing data is the production of maps that can generalize well the processes occurring at the Earth's surface. In this chapter, we discussed the efforts of the community to design automated procedures to produce such maps through image segmentation. Different tasks have been identified and the major trends (land-cover classification, change detection and target/anomaly detection) have been compared experimentally. All these tasks have proven maturity in terms of model design and performance on real datasets. However, the field still needs for efficient implementations, improvements in semisupervised and active learning, and to prepare for domain adaptation problems in the upcoming collection of huge multitemporal sequences. This necessity becomes even more critical when dealing with new generation VHR of hyperspectral imagery, and the new hyperspectral sensors and sounders with thousands of spectral channels. To respond to these needs, we have presented new challenging frameworks such as active learning, semisupervised learning or adaptation models. Contrarily to the standard tasks, these frameworks are still in their infancy for remote sensing applications, thus opening large room for improvement and future developments.

[1] In the remote sensing community, the term 'classification' is often preferred instead of the (perhaps more appropriate) 'segmentation', which is typically used in machine learning and computer vision.

[2] We should note that *labeling* typically involves conducting a terrestrial campaign at the same time the satellite overpasses the area of interest, or the cheaper (but still time-consuming) image photointerpretation by human operators.

[3] http://www.esa.int/esaLP/SEMZHM0DU8E_LPgmes_0.html

[4] http://www.nasa.gov/mission_pages/a-train/a-train.html

CHAPTER 5

Spectral Mixture Analysis

This chapter reviews the distinct field of *spectral mixture analysis* of remote sensing images. The problem consists on extracting spectrally pure pixels directly from the images. This extracted *endmembers* are very useful for image analysis and characterization.

5.1 INTRODUCTION

As we have seen throughout the book and summarized in Chapter 1, hyperspectral sensors acquire the electromagnetic energy in a high number of spectral channels under an instantaneous field of view. These rich datacubes allow us identifying materials in the scenes with high detail. However, high spectral resolution is achieved at the price of spatial resolution due to technical constraints, and the resulting pixels are spatially coarse. A direct consequence is that the spectral vectors acquired are no longer pure but rather mixtures of the spectral signatures of the materials present in the scene. In addition, one could argue that, mixed pixels always exist as they are fundamentally due to the heterogeneity of the landscape and not only because of the characteristics of the sensor.

In this scenario, only a small fraction of the available pixels can be considered as *pure*, i.e., composed by a single material and thus representative of its true spectral signature. The field of spectral mixture analysis (or *spectral unmixing* for short) is devoted to both identifying the most probable set of pure pixels (called *endmembers*) and estimating their proportions (called *abudances*) in each of the image pixels. In fact, when the endmembers have been identified, every single pixel in the image can be synthesized as a linear (or nonlinear) combination of them. The process of unmixing allows many interesting applications which are mainly related to subpixel target detection [Chang, 2003, Shaw and Manolakis, 2002] (that was revised in Chapter 4), crop and mineral mapping [Keshava and Mustard, 2002], and multitemporal monitoring [Zurita-Milla et al., 2011]. In signal processing terms, the spectral unmixing problem can be casted as a *source separation problem*. Actually, the field has clear and interesting connections to *blind source separation* and *latent variable analysis*. However, spectral unmixing maintain its originality, mainly due to the strong constraints related to atmospheric physics, and the hypothesis of dependence between endmembers (sources), which makes the *direct* application of ICA-like approaches not appropriate unless specific physical constraints are introduced.

5.1.1 SPECTRAL UNMIXING STEPS

Basically, spectral mixture analysis involves three steps, which are illustrated in Fig. 5.1:

a. *Dimensionality reduction*. Spectral unmixing intrinsically asumes that the dimensionality of hyperspectral data is lower and can be expressed in terms of the endmembers (a kind of basis). Some methods requiere a previous dimensionality reduction, either feature selection or extraction. The most common approaches use a physically-based selection of the most information bands or rely on the application of principal component analysis (PCA) or the minimum noise fraction (MNF) transformation. See Chapter 3 for details on feature extraction transforms.

b. *Endmember extraction*. The second step deals with the search of a proper vector basis to describe all the materials in the image. Many approaches exist in the literature but, roughly speaking, they can be divided in two groups: the first one tries to find the most extreme spectra, which are the purest and those better describing the vertices of the simplex; the second looks for the spectra that are the most statistically different.

c. *Abundance estimation*. The third and last step is model inversion that typically exploits linear or nonlinear regression techniques for estimating the mixture of materials, called abudance, in each image pixel. A wide variety of methods, from linear regression to neural networks and support vector regression, is commonly used for this purpose. The inversion step consists in solving a constrained least squares problem which minimizes the residual between the observed spectral vectors and the linear space defined by the endmembers.

Due to the close relationship between the second and third steps, some hyperspectral unmixing approaches implement the endmember determination and inversion steps simultaneously. Nevertheless, the common strategy is to define two separated steps: first the endmember extraction and then the linear or nonlinear abundance estimation. Finally, we should note that in this scheme, we did not include the (mandatory) first steps of geometrical and atmospheric image correction. Spectral unmixing is typically done with reflectance data, even though it could be equally carried out on radiance data, as far as the atmosphere affects equally all pixels in the scene. This is, however, a very strong assuption.

5.1.2 A SURVEY OF APPLICATIONS

Spectral unmixing is one of the most active fields in remote sensing image analysis. Unmixing procedures are particularly relevant in applications where the spectral richness is required, but the spatial resolution of the sensors is not sufficient to respond to the precise needs of the application. This section briefly reviews some fields where unmixing has permitted relevant advances.

Standard mapping applications. Mapping of vegetation and of minerals are typical problems for spectral unmixing [Keshava and Mustard, 2002]. Abundance estimation of vegetation in deserts was applied in [Smith et al., 1990, Sohn and McCoy, 1997]. The abundance maps were readily applicable to classify multispectral images based on the fractions of endmembers: in [Adams et al., 1995] the process served to detect landcover changes in the Brazilian Amazon, while Roberts et al. [1998] used multiple endmember spectral mixture models to map chaparral. Elmore et al. [2000] studied spectral unmixing for quantifying vegetation change in semiarid environments. Goodwin et al. [2005] assessed plantation canopy condition from airborne imagery using spectral mixture analysis via fractional abundance estimation. Pacheco and McNairn [2010] performed a spectral unmixing analysis for crop residue mapping in multispectral images, while [Zhang et al., 2004, 2005a] used spectral unmixing of normalized reflectance data for the deconvolution of lichen and rock mixtures. A few applications in the urban environment can be found in the literature: Wu [2004] used a spectral mixture analysis approach for monitoring urban composition using ETM+ images, while in [Dopido et al., 2011] the authors use linear unmixing techiniques to extract features and then performing supervised urban image classification.

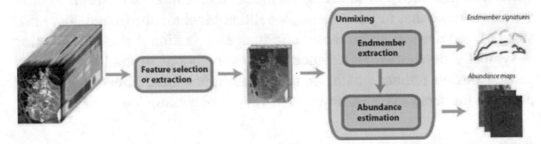

Figure 5.1: Schematic of the hyperspectral unmixing process. First, from a hyperspectral image, a dimensionality reduction step (feature selection or extraction) can be applied to the already geometrically and atmospherically corrected image: while not being strictly necessary, this step is needed by some unmixing methods. In this example, we used the MNF transform (see Chapter 3). Then, the unmixing process consists of (either jointly or separately) the determination of the purest spectra in the image (the endmembers signatures), and to retrieve abundance maps for each one of them.

Multitemporal studies. In the last decade, simultaneously to the development of advanced unmixing models, there has been a high increase in specific applications, not only attached to analyze particular areas, but also to multitemporal monitoring of areas of interest. For example, an important issue is the assessment of the temporal evolution of the covers. In [Shoshany and Svoray, 2002], a multi-date adaptive unmixing was applied to analyze ecosystem transitions along a climatic gradient. Lobell and Asner

[2004] inferred cropland distributions from temporal unmixing of MODIS data. Recently, in [Zurita-Milla et al., 2011], a multitemporal unmixing of medium spatial resolution images was conducted for landcover mapping.

Multisource models. As described above, spectral unmixing was first studied to respond to a lack in spatial detail in multi- and hyperspectral images. The spatial resolution available in the 1990s was very coarse, so efforts on combining different resolution images and ancilliary data were conducted: Puyou-Lascassies et al. [1994] validated a multiple linear regression as a tool for unmixing coarse spatial resolution images acquired by AVHRR. García-Haro et al. [1996] proposed an alternative approach which appends the high spatial resolution image to the hyperspectral data and computes a mixture model based on the joint data set. The method successfully modeled the vegetation amount from optical spectral data. Combining the spectral richness of hyperspectral images with more recent (very) high resolution images also opened new possibilities to perform multisource analysis: fusing information from sensors with different spatial resolution was tackled in [Zhukov et al., 1999], while Small [2003] considered fusion problems with very high resolution imagery. Recently, Amorós-López et al. [2011] exploited a spatial unmixing technique to obtain a composite image with the spectral and temporal characteristics of the medium spatial resolution image and the spatial detail of the high spatial resolution image. A temporal series of Landsat/TM and ENVISAT/MERIS FR images illustrated the potential of the method.

5.1.3 OUTLINE

The remainder of the chapter reviews the main signal processing approaches in the field. Section 5.2 pays attention to the assumption of linearity of the mixture and sets the mixing model in a formal mathematical framework. Section 5.3 reviews the main methods to identify a reasonable number of endmembers in the scene. Section 5.4 summarizes the existing approaches to determine the most pure pixels in the scene, along with the physical or mathematical assumptions used. Section 5.5 details the inversion algorithms to estimate the abundance of each pure constituent in the mixed pixels. Section 5.6 gives some final conclusions and remarks.

5.2 MIXING MODELS

5.2.1 LINEAR AND NONLINEAR MIXING MODELS

The assumption of mixture linearity, though mathematically convenient, depends on the spatial scale of the mixture and on the geometry of the scene [Keshava and Mustard, 2002]. Linear mixing is a fair assumption when the mixing scale is macroscopic [Singer and McCord, 1979] and when the incident light interacts with only one material [Hapke, 1981]. See Fig. 5.2 for an illustrative diagram: the acquired spectra are considered as a linear combination of the endmember signatures present in the scene, $\mathbf{m}_i$, weighted by the respective fractional abundances α_i. The simplicity of such model has given rise to many algorithmical developments and applications.

On the contrary, nonlinear mixing holds when the light suffers multiple scattering or interfering paths, which implies that the acquired energy by the sensor results from the interaction with many different materials at different levels or layers [Borel and Gerstl, 1994]. Figure 5.3 illustrates the two more habitual nonlinear mixing scenarios: the left panel represents an *intimate mixture model*, in which the different materials are close, while the right panel illustrates a *multilayered mixture model*, where interactions with canopies and atmosphere happen sequentially or simultaneously.

The rest of this chapter will mainly focus on the linear mixing model. On the one hand, this is motivated by: 1) it is simple but quite effective in many real settings; 2) it constitutes an acceptable approximation of the light scattering mechanisms; and 3) its formulation is of mathematical convenience and inspires very effective and intuitive unmixing algorithms. On the other hand, nonlinear mixing models are more complex, computationally demanding, mathematically intractable, and highly dependent of the particular scene under analysis. In any case, some nonlinear effective methods are available, such as the two-stream method to include multiple scattering in BRDF models [Hapke, 1981], or radiative transfer models (RTM) that describe the transmitted and reflected radiation for different soil and vegetation covers [Borel and Gerstl, 1994]. As we will see later in the chapter, difficulties related to nonlinear mixture analysis can be circumvented by including nonlinear regression methods in the last step of abundance estimation. For a more detailed review of the literature on unmixing models, we address the reader to [Bioucas-Dias and Plaza, 2010, Plaza et al., 2011] and references therein.

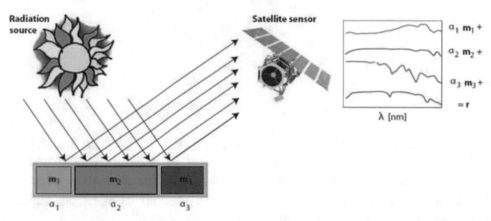

Figure 5.2: Illustration of the spectral linear mixing process. A given material is assumed to be constituted at a subpixel level by patches of distinct materials $\mathbf{m}_i$ contributing linearly through a set of weights or abundances α_i to the acquired reflectance $\mathbf{r}$.

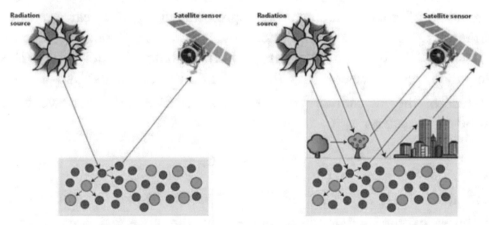

Figure 5.3: Two nonlinear mixing scenarios: the intimate mixture model (left) and the multilayered mixture model (right).

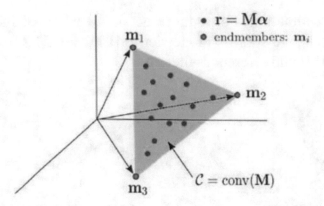

Figure 5.4: Illustration of the simplex set $\mathcal{C}$ for $p = 3$. Points in red denote the available spectral vectors $\mathbf{r}$ that can be expressed as a linear combination of the *endmembers* $\mathbf{m}_i$, $i = 1, \ldots, 3$, (vertices circled in green). The subspace formed defined by these endmembers is the convex hull $\mathcal{C}$ of the columns of $\mathbf{M}$. Figure adapted from [Bioucas-Dias and Plaza, 2010].

5.2.2 THE LINEAR MIXING MODEL

When multiple scattering can be reasonably disregarded, the spectral reflectance of each pixel can be approximated by a linear mixture of endmember reflectances weighted by their corresponding fractional abundances [Keshava and Mustard, 2002]. Notationally, let $\mathbf{r}$ be a $B \times 1$ reflectance vector, where B is the total number of bands, and $\mathbf{m}_i$ is the signature of the ith endmember, $i = 1, \ldots, p$. The reflectance vector can then be expressed as

$$\mathbf{r} = \mathbf{M}\alpha + \mathbf{n},$$

where $\mathbf{M} = [\mathbf{m}_1, \mathbf{m}_2, \ldots, \mathbf{m}_p]$ is the *mixing matrix* and contains the signatures of the endmembers present in the observed area, $\alpha = [\alpha_1, \alpha_2, \ldots, \alpha_p]^\mathsf{T}$ is the fractional abundance vector, and $\mathbf{n} = [n_1, \ldots, n_B]^\mathsf{T}$ models additive noise in each spectral channel. Given a set of reflectances $\mathbf{r}$, the problem basically reduces to estimate appropriate values for both $\mathbf{M}$ and α. This estimation problem is usually completed with two physically reasonable constraints: 1) all abundances must be positive, $\alpha_i \geq 0$, and 2) they have to sum one, $\sum_{i=1}^{p} \alpha_i = 1$, since we want a plausible description of the mixture components for each pixel in the image.

Now, assuming that the columns of $\mathbf{M}$ are independent, the set of reflectances fulfilling these conditions form a $(p-1)$-simplex in $\mathbb{R}^B$. See an illustrative example in Fig. 5.4, showing the simplex set $\mathcal{C}$ for an hypothetical mixing matrix $\mathbf{M}$ containing three endmembers ($\mathcal{C}$ is the *convex hull* of the columns of $\mathbf{M}$). Points in red denote spectral vectors, whereas the vertices of the simplex (in green) correspond to the endmembers. Inferring the mixing matrix $\mathbf{M}$ may be seen as a purely geometrical problem that reduces to identify the vertices of the simplex $\mathcal{C}$.

5.3 ESTIMATION OF THE NUMBER OF ENDMEMBERS

The first step in the spectral unmixing analysis tries to estimate the number of endmembers present in the scene. It is commonly accepted that such number is lower than the number of bands, B. This mathematical requirement means that the spectral vectors lie in a low-dimensional linear subspace. The identification of such dimensionality would not only reveal the intrinsic dimensionality of the data but also would reduce the computational complexity of the unmixing algorithms because data could be projected onto this low-dimensional subspace. Many methods have been applied to estimate the intrinsic dimensionality of the subspace.

Most of the methods involve solving eigenproblems. While principal component analysis (PCA) [Jollife, 1986] looks for projection of data that summarize the variance of the data, the minimum noise fraction (MNF) [Green et al., 1988a, Lee et al., 1990] seeks for the projection that optimizes the signal-to-noise ratio (see Chapter 3 for details). Recently, the hyperspectral signal identification by minimum error (HySime) method [Bioucas-Dias and Nascimento, 2005] is a very efficient method to estimate the signal subspace in hyperspectral images. The method also solves an eigendecomposition problem and no parameters must be adjusted, which makes it very convenient for the user. Essentially, the method estimates the best signal and noise eigenvectors that represent the subspace in MSE terms.

Information-theoretic approaches also exist: from independent component analysis (ICA) [Lennon et al., 2001, Wang and Chang, 2006a] to projection pursuit [Bachmann and Donato, 2000, Ifarraguerri and Chang, 2000]. The identification of the signal subspace has been also tackled with general-purpose subspace identification criteria, such as the minimum description length (MDL) [Schwarz, 1978] or the Akaike information criterion (AIC) [Akaike, 1974]. It is worth noting the work of Harsanyi et al. [1993], in which a Neyman-Pearson detection method (called HFC) was successfully introduced and evaluated. The method determines the number of spectral endmembers in hyperspectral data and implements the so-called virtual dimensionality (VD) [Chang and Du, 2004], which is defined as the minimum number of spectrally distinct signal sources that characterize the hyperspectral data. Essentially, the estimated dimensionality p by HFC reduces to the highest number for which the correlation matrix have smaller eigenvalues than the covariance matrix. A modified version of HFC is the noise-whitened HFC (NWHFC), which includes a noise-whitening process as preprocessing step to remove the second-order noise statistical correlation [Chang and Du, 2004].

All the methods considered so far only account for second-order statistics with the exception of some scarce application of ICA. Nonlinear manifold learning methods that describe local statistics on the image probability density function (PDF) have been applied to describe the intrinsic dimensionality, such as ISOMAP and locally linear ambedding (LLE) [Bachmann et al., 2005, 2006, Gillis et al., 2005, Yangchi et al., 2005]. All these methods were revised in Chapter 3.

5.3.1 A COMPARATIVE ANALYSIS OF SIGNAL SUBSPACE ALGORITHMS

The hyperspectral scene used in the experiments is the AVIRIS Cuprite reflectance data set[1]. The data was collected by the AVIRIS spectrometer over the Cuprite Mining area in Nevada (USA) in 1997. This scene has been used to validate the performance of many spectral unmixing and abundance estimation algorithms.

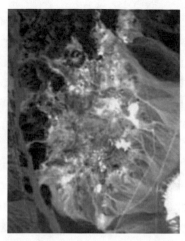

Figure 5.5: RGB composition of a hypespectral AVIRIS image over a mine area in Nevada, USA.

A portion of 290×191 pixels is used here (see an RGB composition in Fig. 5.5). The 224 AVIRIS spectral bands are between 0.4 and 2.5μm, with a spectral resolution of 10 nm. Before the analysis, several bands were removed due to water absorption and low SNR, retaining a total of 188 bands for the experiments.

The Cuprite site is well understood mineralogically, and has several exposed minerals of interest as reported by the U.S. Geological Survey (USGS) in the form of various mineral spectral libraries[2] used to assess endmember signature purity (see Fig. 5.6). Even though the map was generated in 1995 and the image was acquired in 1997, it is reasonable to consider that the main minerals and mixtures are still present. This map clearly shows several areas made up of pure mineral signatures, such as buddingtonite and calcite minerals, and also spatially homogeneous areas made up of alunite, kaolinite, and montmorillonite at both sides of the road. Additional minerals are present in the area, including chalcedony, dickite, halloysite, andradite, dumortierite, and sphene. Most mixed pixels in the scene consist of alunite, kaolinite, and muscovite. The reflectance image was atmospherically corrected before confronting the results with the signature library.

In order to estimate an appropriate number of endmembers in the scene, we used some representative methods: PCA, MNF, HFC, NWHFC and HySime. Figure 5.7 shows the results obtained analyzing the eigenvalues through PCA and the MNF transforms. The left plot presents the cumulative explained variance (energy) as a function of the number of eigenvalues. We can see that the spectral signal energy contained in the first eight eigenvalues is higher than 99.95% of the total explained variance. Unlike PCA, the MNF transform yields a higher number of distinct pure pixels, $p = 13$ (see Fig. 5.7[middle]). Figure 5.7[right] summarizes the results obtained by the HySime method, which estimates $p = 18$. The figure shows the evolution of the MSE as a function of the parameter k. The minimum of the MSE occurs at $k = 18$, which corresponds to the estimated number of endmembers present in the image. As expected, the projection of the error and noise powers display decreasing and increasing behaviors, respectively, as a function of the subspace dimension k. The VD for both HFC and NWHFC was estimated with the false-alarm probability set to different values $P_f = \{10^{-2}, \ldots, 10^{-6}\}$ (see Table 5.1). Previous works fixed a reasonable value of $P_f = 10^{-5}$, which gives rise to p around 14. Since these results agree with [Nascimento and Bioucas-Dias, 2005a], we fixed the number of endmembers $p = 14$ in the experiments of the following sections.

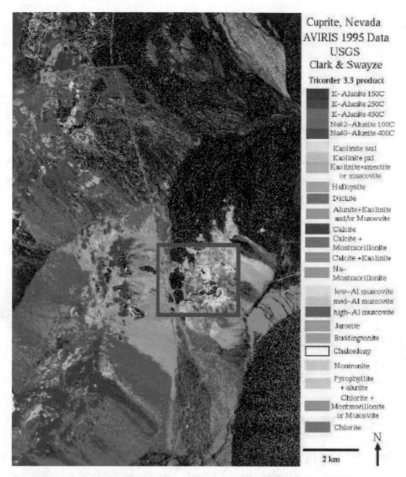

Figure 5.6: USGS map showing the location of different minerals in the Cuprite mining site. The box indicates the study area in this chapter. The map is available online at `http://speclab.cr.usgs.gov/cuprite95.tgif.2.2um_map.gif`

5.4 ENDMEMBER EXTRACTION

This section reviews the different families of endmember extraction techniques. Issues related to representativeness and variability of endmembers signatures are also discussed.

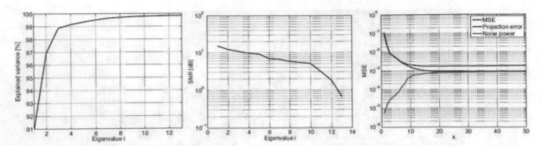

Figure 5.7: Analysis of the intrinsic dimensionality of the Cuprite scene by means of principal component analysis (PCA), minimum noise fraction (MNF), and the HySime method (right).

Table 5.1: VD estimates for the Aviris Cuprite scene with various false alarm probabilities [Plaza and Chang, 2006].

Method	P_F				
	10^{-2}	10^{-3}	10^{-4}	10^{-5}	10^{-6}
HFC [Harsanyi et al., 1993]	23	20	17	16	**14**
NWHFC [Chang and Du, 2004]	21	18	16	**14**	12
HySime [Bioucas-Dias and Nascimento, 2005]			18		

5.4.1 EXTRACTION TECHNIQUES

The previous step yields an estimated number of endmembers present in the scene by only looking at the statistical characteristics of the image. This number, however, could be known *a priori* in some specific cases where the area is well characterized. In those cases, manual methods for endmember extraction have been developed [Bateson and Curtiss, 1996, Tompkins et al., 1997]. However, this is not the case in many problems and hence automatic methods for *endmember extraction* are necessary. The problem can be casted as a *blind source separation* one because the sources and the mixing matrix are unknown. Even though ICA has been used to blindly unmix hyperspectral data [Bayliss et al., 1997, Chen and Zhang, 1999, Tu, 2000], its suitability to the problem characteristics has been questioned [Keshava et al., 2000, Nascimento and Bioucas-Dias, 2005b]. The main reason is that the assumption of mutual independence of the sources is not met by the problem in which abundances must sum to one. To solve the problems of ICA, many methods have been introduced. Two main families are identified: *geometrical* and *statistical* approaches [Parente and Plaza, 2010].

The family of geometrical methods consider the properties of the convex hull, and can be split into two subfamilies. The *pure-pixel methods* assume the presence of at least one pure pixel per endmember. This implies that there is at least one spectral vector on each vertex of the data simplex. This assumption limits the generality of the approaches but leads to very high computational efficiency. On the contrary, the family of *minimum-volume methods* estimate the mixing matrix $\mathbf{M}$ by minimizing the volume of the simplex. This is a nonconvex optimization problem much harder to solve. Geometrical methods can yield poor results when spectra are highly mixed. This is due to the lack of enough spectral vectors in the simplex facets.

Statistical methods can alleviate the above problems of geometrical methods, and constitute a very competitive alternative. In most of the cases, however, statistical methods have more computational burden. This family can be split into 1) methods based on information theory, such as ICA or the Dependent Component Analysis method (DECA), which assumes the abundance values are drawn from a Dirichlet distribution; 2) methods with inspiration in machine learning, such as the one-class SVM classifier that embraces all input samples in a feature spaces implicitly accessible via kernel functions (see Chapter 4) or associative neural networks; and 3) those that impose sparseness in the solution, such as the basis-projection denoising (BPDN) or the iterative spectral mixture analysis (ISMA).

Recently, a third approach based on sparse regression has been introduced [Iordache et al., 2011]. The main existing methods in all these families are summarized in Table 5.2. A more detailed description of the methods can be obtained from [Bioucas-Dias and Plaza, 2010, Plaza et al., 2011] and references therein.

5.4.2 A NOTE ON THE VARIBILITY OF ENDMEMBERS

A last important point to discuss regards the variability of endmembers. Spectral unmixing may give rise to unrealistic results because, even if algorithms succeed, the spectra representing the endmembers might not account for the spectral variability present in the scene. The issue of incorporating endmember variability in the spectral mixture analysis has been long studied and constitutes a crucial issue for the success of abundance estimation procedures [Bateson et al., 2000, García-Haro et al., 2005, Roberts et al., 1998]. Song [2005] reviewed how to incorporate endmember variability in the unmixing process for subpixel vegetation fractions in the urban environment. In [Settle, 2006] the effect of variable endmember spectra in the linear mixture model was studied. Selection of a meaningful set of free parameters and the inclusion of spatial information in the unmixing process have demonstrated useful for stabilizing the solution [García-Vílchez et al., 2011, Plaza and Chang, 2006].

5.4.3 A COMPARATIVE ANALYSIS OF ENDMEMBER EXTRACTION ALGORITHMS

This section compares several endmember determination methods representing different families: 1) pure-pixel geometrical approaches (IEA, VCA, PPI and NFINDR); 2) SISAL for geometrical minimum volume approaches; 3) ICA for the information-theoretic-based methods; and 4) SVDD target detection (see Section 4.4) and EIHA for the machine learning based approaches. To evaluate the goodness-of-fit of the extracted endmembers, we use different scores: the relative RMSE, the spectral angle mapper (SAM), and the spectral information divergence (SID). The SAM between $\hat{m}$ and $\mathbf{m}$ is expressed as

Table 5.2: Taxonomy of endmember extraction methods.

Family	Method	Brief description	Automatic	CPU cost
Geometrical (pure-pixel)	IEA [Neville et al., 1999]	Iteratively selects endmembers that minimize error in the unmixed image	✓	High
	VCA [Nascimento and Bioucas-Dias, 2005a]	Iteratively projects data onto a direction orthogonal to the subspace spanned by the previous endmember	✓	Low
	PPI [Boardman, 1993]	Projects spectra onto many random vectors, stores the most extreme distances to select endmembers	✓	Medium
	N-FINDR [Winter, 1999]	Finds the pixels defining the simplex with maximum volume trough inflation	✓	Medium
	SGA [Chang et al., 2006]	Iteratively grows a simplex by finding the vertices that yield maximum volume	✓	Medium
	SMACC [Gruninger et al., 2004]	Iteratively incorporates new endmembers by growing a convex cone representing the data	×	High
Geometrical (min-volume)	SISAL [Bioucas-Dias, 2009]	Robust version of min-volume by allowing violation of the positivity constraint	✓	Low
	CCA [Ifarraguerri and Chang, 1999]	Iteratively selects endmembers maximizing the correlation matrix eigenspectrum, forces positive endmembers	✓	Low
	MVES [Chan et al., 2009]	Implements a cyclic minimization of a series of linear programming problems to find the min-vol simplex	✓	Low
	MVT-NMF [Miao and Qi, 2007]	Minimizes a regularized problem: a term minimizing the approximation error of NMF and another constraining the volume of the simplex	×	Low
	ICE [Berman et al., 2004]	Similar to MVT-NMF but replaces the volume by sum of squared distances between all simplex vertices	×	Medium
	SPICE (also sparse) [Zare and Gader, 2007]	Extension of ICE with sparsity-promoting prior	×	Medium
	ORASIS [Bowles et al., 1997]	Iterative procedure with several modules involving endmember selection, unmixing, spectral libraries and spatial postprocessing	×	High
Statistical (info. theory)	ICA [Bayliss et al., 1997]	Standard formulation to retrieve the mixing matrix M. Assumes independence of the sources	✓	Low
	DECA [Nascimento and Bioucas-Dias, 2007]	Forces a mixture of Dirichlet densities as prior for the abundance fractions	✓	Low
Statistical (machine learning)	SVDD (also geometrical) [Broadwater et al., 2009]	Hypersphere in kernel space, rejection ratio set to zero	✓	Low
	EIHA [Graña et al., 2009]	Lattice auto-associative memories + image segmentation	✓	High
Statistical (sparse models)	BP/OMP [Pati et al., 2003]	Greedy algorithm based on orthogonal basis pursuit	✓	Low
	BPDN [Chen et al., 2001]	Basis pursuit algorithm with a relaxation to solve the BP/OMP problem	✓	Low
	ISMA [Rogge et al., 2006]	Iteratively finds an optimal endmembers set by examining the change in RMSE after reconstructing the original scene using the estimated fractional abundance	✓	Low

$$\mathrm{SAM} = \mathrm{acos}\left(\frac{\hat{\mathbf{m}}^{\top}\mathbf{m}}{\|\hat{\mathbf{m}}\|\|\mathbf{m}\|}\right).$$

The SID is an information-theoretic measure. Note that the probability distribution vector associated with each endmember signature is given by $\mathbf{p} = \mathbf{m}/\Sigma_i\, m_i$. Then, the similarity can be measured by the (symmetrized) relative entropy

$$D(\hat{\mathbf{m}}, \mathbf{m}) = \sum_i p_i \log\left(\frac{p_i}{\hat{p}_i}\right) + \sum_i \hat{p}_i \log\left(\frac{\hat{p}_i}{p_i}\right).$$

Following the results of the previous section, 14 endmembers were extracted by the different techniques. Then, distances were computed between the endmembers extracted by each algorithm $\hat{\mathbf{m}}$ and the most similar ones in the USGS database $\mathbf{m}$. This spectral library[3] contains 501 cataloged spectra, so the search is a complex problem full of local minima. The signatures provided by each method were scaled by a positive constant to minimize the MSE between them and the corresponding library spectra for a proper comparison.

Figure 5.8 summarizes the performance of the methods in terms of computational cost, relative RMSE, SAM and SID angular errors. Regarding computational effort, the IEA, PPI and EIHA methods involve high computational burden. The rest of the methods are affordable for any current personal computer[4]. Regarding distance of the endmembers to those found in the database, all the methods achieve a relative RMSE lower than 0.2, except for PPI and ICA. Similar trends are observed for SAM and SID. The VCA outperformed the rest of the methods in accuracy (in all measures), and showed very good computational efficiency, closely followed by SVDD, SISAL and IEA. Particularly interesting is the case of the poor performance of ICA in terms of endmember extraction. The method is based on the assumption of mutually independent sources (abundance fractions), which is not the case of hyperspectral data since the sum of abundance fractions is constant, implying statistical dependence among them. Then, even if a basis is successfully extracted, the extracted endmembers by ICA do not necessarily have a physical (os spectra-like) shape because the problem is unconstrained. This is the main reason for the poor numerical results. Nevertheless, when used for synthesis, ICA may yield good results in terms of abundance estimation. More discussion on this issue can be obtained from [Nascimento and Bioucas-Dias, 2005b]. Another interesting issue is the good performance of the SVDD kernel method: in this case we fixed the rejection ratio to 0.001 (intuitively this reduces to embrace all data points in the kernel feature space) and the σ parameter for the RBF's kernel was varied to give the requested $p = 14$. The method can be safely considered the best compromise between accuracy and computational cost among the statistical family, and constitutes a serious competitor to the pure geometrical methods. Actually, the SVDD can be considered a geometrical method with more flexibility to define the convex hull.

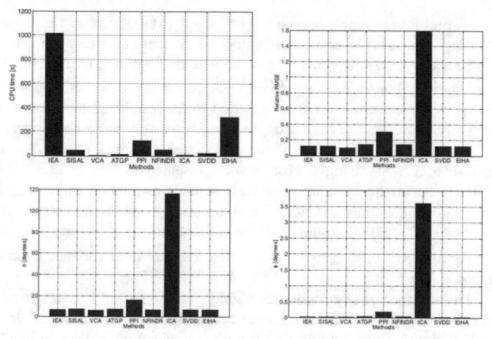

Figure 5.8: Computational cost (in secs.) and accuracy in terms of RMSE, SAM and SID for all the tested endmember extraction methods.

Looking at Fig. 5.9, the estimated signatures are close to the laboratory spectra. The bigger mismatches occur for ♯325 and ♯402 signatures for IEA and the SVDD, but only on particular spectral channels.

5.5 ALGORITHMS FOR ABUNDANCE ESTIMATION

The last step in spectral unmixing analysis consists in estimating the fractional abundances α. In the case that the basis is identified, the problem reduces to model inversion or regression approaches. Other approaches consider the steps of unmixing and abundance estimation jointly. We here review linear or nonlinear approaches found in the literature.

5.5.1 LINEAR APPROACHES

The class of inversion algorithms based on minimizing squared-error start from the simplest form of least squares inversion and increases in complexity as more constraints are included. The unconstrained least-squares problem is simply solved by $\hat{\alpha} = \mathbf{M}^\dagger \mathbf{r} = (\mathbf{M}^\mathrm{T}\mathbf{M})^{-1}\mathbf{M}^\mathrm{T}\mathbf{r}$. The issue of including the sum-to-one constraint means that the LS problem is constrained by $\Sigma \alpha_i = 1$, which can be solved via Lagrange multipliers. The non-negativity constraint is not as easy to address in closed-form as the full additivity. Nevertheless, a typical solution to the problem simply estimates iteratively $\hat{\alpha}$ and, at every iteration, finds a LS for only those coefficients of $\hat{\alpha}$ that are positive using only the associated columns of $\mathbf{M}$. A fully constrained LS unmixing approach was presented in [Heinz and Chang, 2000]. The statistical analogue of least squares estimation minimizes the variance of the estimator. Under the assumption of additive noise, $\mathbf{n}$, with covariance, $\mathbf{C}_n$, the minimum variance estimate of the abundances reduces to $\hat{\alpha} = (\mathbf{M}^\mathrm{T}\mathbf{C}_n^{-1}\mathbf{M})^{-1}\mathbf{M}^\mathrm{T}\mathbf{C}_n^{-1}\mathbf{r}$. This is called the minimum variance unbiased estimator (MVUE).

The number of algorithms for abundance estimation is ever-growing, and many variants and subfamilies are continuously being developed. In [Li and Narayanan, 2004], the use of features extracted after discrete wavelet transform improved the least squares estimation of endmember abundances. Abundance estimation of spectrally similar minerals by using derivative spectra in simulated annealing was presented in [Debba et al., 2006]. In [Chang and Ji, 2006], a weighted abundance-constrained linear spectral mixture analysis was successfully presented.

Miao et al. [2007] introduced the maximum entropy principle for mixed-pixel decomposition from a geometric point of view, and demonstrated that when the given data present strong noise or when the endmember signatures are close to each other, the proposed method has the potential of providing more accurate estimates than the popular least-squares methods. An updated revision of the available methods can be found in [Bioucas-Dias and Plaza, 2010].

5.5.2 NONLINEAR INVERSION

Due to the difficulties posed when trying to model the light interactions, the linear model has been widely adopted. Nevertheless, the model can be completed by plugging a nonlinear inversion method based on the estimated endmembers, either by exploiting the intimate spectral mixture to perform spectral unmixing or by definition of nonlinear decision classifiers. In this latter approach, multilayer perceptrons [Atkinson et al., 1997], nearest neighbor classifiers [Schowengerdt, 1997], and support vector machines (SVMs) [Brown et al., 2000] have been used. Actually, the nonlinear extensions via kernels allow SVMs to deal with overlapping pure pixels distributions and nonlinear mixture regions. Adopting a linear model while performing nonlinear decision boundaries can be done in a feature space via a nonlinear reproducing kernel function [Camps-Valls and Bruzzone, 2009]. See Chapters 4 and 4 for discussion on kernel nonlinear feature extraction and classification, respectively. Following this idea, nonnegative constrained least squares and fully constrained least squares in [Harsanyi and Chang, 1994] were extended to their kernel-based counterparts [Broadwater et al., 2009]. A simple algorithm for doing nonlinear regression with kernels consists of iterating the equations:

$$\hat{\alpha} = (K(\mathbf{M}, \mathbf{M}))^{-1} [K(\mathbf{M}, \mathbf{r}) - \lambda]$$
$$\lambda = K(\mathbf{M}, \mathbf{r}) - K(\mathbf{M}, \mathbf{M})\hat{\alpha},$$

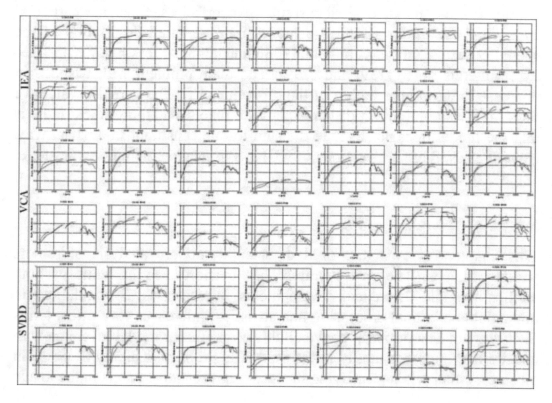

Figure 5.9: Comparison of the extracted signatures (blue) with those in the USGS spectral library (red) for a representative method of each family: IEA (geometrical, pure-pixel), VCA (geometrical, minimum-volume) and SVDD (statistical).

where λ is the Lagrange multiplier vector used to impose the non-negativity constraints of the estimated abundances. Nevertheless, the method does not incorporate the sum-to-one constraint. We refer to this method as the Kernel NonNegativity Least Squares (KNNLS).

5.5.3 A COMPARATIVE ANALYSIS OF ABUNDANCE ESTIMATION ALGORITHMS

We here compare LS and KNNLS methods for abundance estimation. For the sake of brevity, we only apply the methods to the endmembers identified by the VCA method, which gave the best overall results. Figure 5.10 illustrates the estimated abundance maps. The obtained maps nicely resemble the available geological maps. In general, both linear and nonlinear methods yield similar results. Nevertheless, for some particular cases, the use of the KNNLS achieves more detailed description of the spatial coverage (see e.g., minerals ♮386 and ♮245) or less noisy maps (see e.g., minerals ♮126, ♮139, and ♮299). The main problem with KNNLS in this case is the proper tuning of the σ parameter for the kernels, which we fixed to the sample mean distance between pixels. Another critical issue is that the model does not include the sum-to-one constraint in a trivial way. Nevertheless, there are versions of KNNLS to include them [Broadwater et al., 2009].

5.6 SUMMARY

This chapter reviewed the field of spectral mixture analysis of remote sensing images. The vast literature has been revised in terms of both applications and theoretical developments. We observed tight relations to other fields of signal processing and statistical inference: very often, terms like 'domain description', 'convex hull', or 'nonlinear model inversion with constraints' appear naturally in the field. The problem is highly ill-posed in all steps, either when identifying the most representative pixels in the image or when inverting the mixing model. These issues pose challenging problems. Currently, quite efficient methods are available, which has led to many real applications, mainly focused on identifying crops and minerals to finally deploy abundance (probability) maps. All steps in the processing chain were analyzed and motivated theoretically and also tested experimentally.

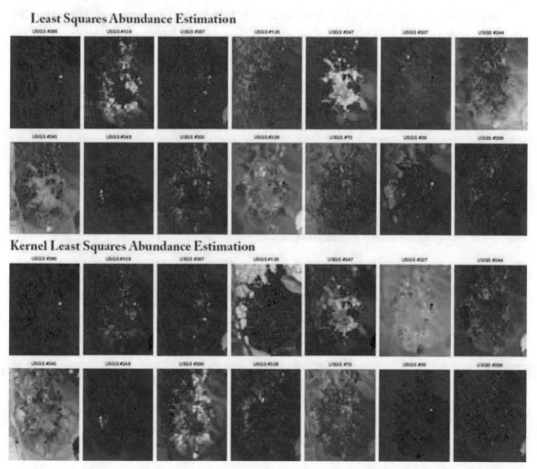

Figure 5.10: Abundance maps of different minerals estimated using least squares linear unmixing (top panel) and the kernel version (bottom panel) for the VCA method. The different materials identifiers in the USGS report are indicated above.

[2] http://speclab.cr.usgs.gov/spectral-lib.html

[3] Very often we are aware of the type of class we are looking for. In this case, we may have some labeled pixels of the class of interest or we can rely on databases (often called *spectral libraries*) of class-specific spectral signatures. In this case, the problem reduces to defining the background class and then detecting spectral signatures that are closer to the known signatures than to the background.

[4] All experiments were carried in Matlab usinga2GbRAM,2 processors, 3.3 GHz CPU station.

CHAPTER 6

Estimation of Physical Parameters

This chapter revises the main problems and methods used in the field of model inversion and estimation of physical parameters from remotely sensed images. The number and accuracy of current methods (either physically-based inversion, statistical methods or combined approaches) has made possible the prediction of atmospheric parameters, crop characteristics, and ocean state.

6.1 INTRODUCTION AND PRINCIPLES

The main goal of optical Earth Observation is to monitor the Earth and its interaction with the atmosphere. The analysis can be done at local or global scales by looking at bio-geo-chemical cycles, atmospheric situations and vegetation dynamics [Liang, 2004, 2008, Lillesand et al., 2008, Rodgers, 2000]. All these complex interactions can be studied through the definition of biophysical parameters either representing different properties for land (e.g., surface temperature, crop yield, defoliation, biomass, leaf area coverage), water (e.g., yellow substance, ocean color, suspended matter or chlorophyll concentration) or the atmosphere (e.g., temperature and moisture profiles at different altitudes). Every single application should consider the specific knowledge about the physical, chemical and biological processes involved, such as energy balance, evapotranspiration, or photosynthesis. However, remotely-sensed observations only sample the radiation reflected or emitted by the surface (see Chapter 1) and thus, an intermediate modeling step is necessary to transform the measurements into estimations of the biophysical parameters [Baret and Buis, 2008].

This chapter reviews a distinct and very important field of remote sensing data processing: that concerned with the estimation of bio-geophysical parameters from images and ancillary data. The acquired data may consist of multispectral or hyperspectral images provided by satellite or airborne sensors, but can also integrate alternative measurements: spectra acquired by *in situ* (field) radiometers, GIS data that help to integrate geographical information, or radiosonde measures for weather monitoring. A series of international study projections, such as the International Geosphere-Biosphere Programme (IGBP), the World Climate Research Programme (WCRP), and the NASA's Earth Observing System (EOS), established remote sensing model inversion as the most important problem to be solved with EO imagery in the near future. Monitoring the physical processes occurring in land, ocean and the atmosphere is now possible and reliable due to the improved image characteristics (increasing spatial, spectral, multiangular and temporal resolutions) and the advances in signal processing and machine learning tools.

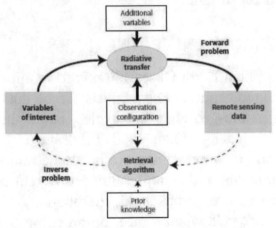

Figure 6.1: Forward (solid lines) and inverse (dashed lines) problems in remote sensing. Figure adapted from [Baret and Buis, 2008].

6.1.1 FORWARD AND INVERSE MODELING

Sensing an object remotely implies that the system is observed in an indirect and complicated way. For example, the presence of the atmosphere between the object and the imaging system introduces nonlinear effects on the acquired signals. Also, the sensors are commonly affected by noise and distortions, and high levels of uncertainty can be involved in the measurement. This is why very often the problem is formalized in terms of inverting a complex function, and hence it can be framed in the *Inverse Systems Theory*.

Two directions are considered when analyzing the system, as illustrated in Fig. 6.1. The *forward* (or direct) problem involves *radiative transfer models* (RTMs). These models summarize the physical processes involved in the energy transfer from canopies and atmosphere to measured radiance. They simulate the reflected or emitted radiation transmitted through the atmosphere for a given observation configuration (e.g., wavelength, view and illumination directions) and some auxiliary variables (e.g., vegetation and atmosphere characteristics). Solving the *inversion* problem implies the design of algorithms that, starting from the radiation acquired by the sensor, can give accurate estimates of the variables of interest, thus 'inverting' the RTM. In the inversion process, *a priori* information of the variables of interest can also be included to improve the performance, such as the type of surface, geographical location, acquisition time or statistical properties of the data distribution.

The real physical system that relates the atmosphere and the land surface is very complex. Simple approximations describe the system as a function parametrized by a comprehensive set of weights. Notationally, a *discrete forward model* that describes such a system can be expressed as

$$\mathbf{y} = f(\mathbf{X}, \boldsymbol{\theta}) + \mathbf{n}, \tag{6.1}$$

where $\mathbf{y}$ is a set of measurements (such as the expected radiance); $\mathbf{X}$ is a matrix of state vectors that describe the system (e.g., the parameters such as temperature or moisture); $\boldsymbol{\theta}$ contains a set of controllable measurement conditions (such as different combinations of wavelength, viewing direction, time, Sun position, and polarization); $\mathbf{n}$ is an additive noise vector; and $f(\cdot)$ is a function which relates $\mathbf{X}$ with $\mathbf{y}$. Such function is typically considered to be nonlinear, smooth and continuous. The *discrete inverse model* can be then simply defined as

$$\hat{\mathbf{X}} = g(\mathbf{y}, \boldsymbol{\omega}), \tag{6.2}$$

that is, finding a (possibly nonlinear) function $g(\cdot)$, parametrized by weights $\boldsymbol{\omega}$, that approximates the measurement conditions, $\mathbf{X}$, using a set of observations as inputs, $\mathbf{y}$.

6.1.2 UNDETERMINATION AND ILL-POSED PROBLEMS

In remote sensing, the problem of inverting the function f is in general *undetermined* and highly *ill-posed*: the number of unknowns is generally larger than the number of independent radiometric information remotely sampled by the sensors. This issue has been largely reported in the literature [Combal et al., 2001, Knyazikhin, 1999, Liang, 2008, Verstraete and Pinty, 1996]. In general, estimating physical parameters from either RTMs or satellite-derived data constitutes a very difficult problem due to the presence of high levels of uncertainty, which are primarily associated to atmospheric conditions and sensor calibration, sun angle, and viewing geometry. This translates into inverse problems in which very similar reflectance spectra, $\mathbf{y}$, may correspond to very diverse parameterizations (possible solutions), $\mathbf{X}$. Another existing problem is the generally unknown nature of the noise present in the problem. But probably the most important problem is related to the poor sampling of the parameter space in most of the applications. Similarly, to what discussed for classification in Section 4.5.1, the ratio between the number of samples available and the number of spectral channels is usually very low, leading to ill-posed problems with more *unknowns* than *observations*, when it comes to invert the RTM. In such inverse problems, *regularization* of the solution plays a crucial role. This is typically addressed by including prior information or constraints on the distribution of the variables. For example, imposing a prior distribution on the expected values of the variable to be predicted, or including spatial/spectral/temporal constraints very often alleviates the problem [Liang, 2004].

6.1.3 TAXONOMY OF METHODS AND OUTLINE

Methods for model inversion and parameter retrieval can be roughly separated in three main families: statistical, physical and hybrid methods:

- The *statistical* inversion models are reviewed in Section 6.2. One can identify two types of statistical models: parametric and non-parametric. Parametric models rely on physical knowledge of the problem and build explicit parametrized expressions that relate a few spectral channels with the bio-geo-physical parameter(s) of interest. On the other hand, non-parametric models are adjusted to predict a variable of interest using a training dataset of input-output data pairs. The used data to adjust the models come from concurrent measurements of the variable of interest (e.g., soil moisture) and the corresponding observations (e.g., reflectance or radiances). In this latter case, a terrestrial campaign is necessary at the same time the satellite overpasses the study area to measure the surface parameter. Oceanic applications, such as the estimation of oceanic chlorophyll concentration, involve measuring the reflectance using in-water instrumentation and platforms. For atmospheric studies, it becomes necessary to perform atmospheric radiosonde measurements of, for instance, temperature, water vapor, or ozone concentration. As we will see later, many machine learning approaches have been used in all these problems.

- The second family of methods are known as *physical* inversion models, and try to reverse RTMs. There are several approaches to this complex problem that will be revised in Section 6.3. After generating input-output (parameter-radiance) datasets, the problem reduces to, given new spectra, searching for similar spectra in the dataset and assigning the most plausible ('closest') parameter. By 'searching' we mean to minimize a proper cost or loss functional. The use of RTMs to generate datasets is a common practice, and especially convenient because acquisition campaigns are very costly (in terms of time, money, and human resources) and usually limited in terms of parameter combinations. RTMs are also widely used in the preliminary phase of a new sensor design, which allows understanding both the limits and capabilities of the instrument for the retrieval tasks.

- Lately, the family of *hybrid* inversion models try to combine the previous approaches, and will be revised in Section 6.4. Hybrid methods exploit the input-output data generated by simulations of RTMs, to train regression models (typically neural networks) that invert the model. In this setup, nonparametric regression models are very efficient and can replace more costly physical inversion methods.

6.2 STATISTICAL INVERSION METHODS

This section reviews the state-of-the-art in *statistical* (retrieval or regression-based) methods in the application fields of land, ocean and atmospheric parameter retrieval.

6.2.1 LAND INVERSION MODELS

The estimation of vegetation properties from remote sensing images help to determine the phenological stage and health status (e.g., development, productivity, stress) of crops and forests [Hilker et al., 2008]. Leaf chlorophyll content (*Chl*), leaf area index (LAI), and fractional vegetation cover (FVC) are among the most important vegetation parameters [Lichtenthaler, 1987, Whittaker and Marks, 1975].

Parametric methods

To date, a large number of spectral vegetation indices (VIs) have been developed for the study of *Chl* based on leaf reflectance [Haboudane et al., 2004, Myneni et al., 1995]. These indices are simple parametric combinations of several spectral channels. They are designed to reinforce the sensitivity to particular bio-physical phenomena, such as greenness or water content, while being insensitive to other factors affecting the shape of the spectral signature, such as soil properties, solar illumination, atmospheric conditions, and sensor viewing geometry. Therefore, as pointed out in Section 3.4, spectral indices can be considered as meaningful extracted features that help in the estimation of the bio-physical variables to predict.

Different narrowband VIs have been proposed for the general study of vegetation status, which have proven their predictive power in assessing *Chl* and other leaf pigments, as well as vegetation density parameters like the leaf area index (LAI) and the fractional vegetation cover (FVC) [Berni et al., 2009, Haboudane et al., 2008, Le Maire et al., 2008]. Many of these indices have been derived from high resolution spectrometers, which include many (up to more than two hundred) hyperspectral bands. The simple calculation of these indices has made possible deriving reasonable maps of vegetation properties in a quick and easy way. Furthermore, since the launch of imaging spectrometers into spacecrafts, these VIs have been applied at canopy level on ecosystems across the globe [Stagakis et al., 2010, Verrelst et al., 2008]. Nevertheless, the majority of the indices only use up to five bands. Their simplicity proved to be desirable for numerous mapping applications, but it is also recognized that they under-exploit the full potential of the hyperspectral datacube [Schaepman et al., 2009].

Vegetation indices have been used to predict bio-physical variables from reflectance data $\mathbf{y}$. The inversion models are typically described by one of the following parametric expressions [Liang, 2004]:

$$
\begin{aligned}
x &= \sum_{i=1}^{n} a_i \mathrm{VI}^i \\
x &= a + b\mathrm{VI}^c \\
x &= a \ln(b - \mathrm{VI}) + c,
\end{aligned}
\tag{6.3}
$$

where VI is a combination (typically ratios) of reflectance values in n specific channels. The indices can be either computed using digital numbers, TOA radiance/reflectance, or surface radiance/reflectance.

A key issue to design VIs considers the characteristics of the soil reflectance. Let us illustrate how differences between the visible red and near-infrared (NIR) bands of a Landsat image can be used to identify areas containing significant vegetation. Figure 6.2 shows a standard color-infrared (CIR) composite of the image before and after a decorrelation stretching. This operation emphasizes spectral differences across the scene. Healthy, chlorophyll-rich vegetation has a high reflectance in the NIR (highlighted here in green). By looking at the differences between the NIR and the red channels, it is possible to quantify this contrast in spectral content between vegetated areas and other surfaces such as pavement, bare soil, buildings, or water. The Normalized Difference Vegetation Index (NDVI) [Liang, 2004, Rouse et al., 1973] is the most widely used index in the literature and is defined as:

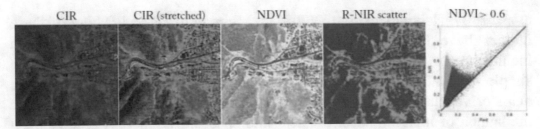

| CIR | CIR (stretched) | NDVI | R-NIR scatter | NDVI> 0.6 |

Figure 6.2: Landsat image acquired over a residential area containing different land classes (asphalt, forest, buildings, grass, water, etc.). Left to right: standard color-infrared (CIR) composite, stretched CIR and NDVI image, thresholded NDVI image, and the scatter plot of all image pixels in Red versus NIR space.

$$NDVI = \frac{NIR - R}{NIR + R}. \tag{6.4}$$

The idea underlying this expression is that the difference between the NIR and the red should be large for high chlorophyll density (see Chapter 3). The normalization simply discounts the effects of uneven illumination, such as shadows of clouds or hills. A common practice to identify significant vegetation in an image consists on applying a simple threshold to the NDVI image, as shown in the fourth panel of Fig. 6.2.

A scatter plot of all image pixels in the space Red-NIR shows a clear positive linear correlation of the soil reflectances. The red line that limits the triangular-shaped distribution is called the *soil line* (in red in the rightmost panel of Fig. 6.2). It is also worth noting that those pixels with higher NIR reflectance and lower red reflectance are dense vegetation canopies. Over a vegetated surface, this line often constitutes the base of a triangle shape that is taken as a normalization factor of many VIs. In this optic, many corrected VIs such as the mSAVI (Improved Soil-Adjusted Vegetation index, [Qi et al., 1994]) and the TSAVI (Transformed Soil-Adjusted Vegetation Index, [Rondeaux et al., 1996]) have been proposed in the inversion literature.

Empirical approaches typically rely on carefully selected band ratios. Many indices focus on the red-edge reflectance, while others rely on describing the triangular-shaped distribution and on the area under the reflectance curve [Liang, 2008]. Section 6.5 shows a quantitative comparison of the most common indices for the particular case of estimating surface chlorophyll content from hyperspectral images. For a comprehensive revision of other indices, see [Liang, 2004, 2008]. Finally, we should note that indices are sensor-specific because of their particular spectral configurations. Figure 6.3[left] shows a vegetation index obtained with AVHRR.

Non-parametric methods

Several nonparametric approaches have been introduced for land parameter retrieval. In [R. and Nichol, 2011], biomass was estimated using common spectral band ratios, vegetation indices and linear/stepwise multiple regression models. Partial least squares (PLS) regression has been used for mapping canopy nitrogen [Coops et al., 2003, Townsend et al., 2003], analyzing biophysical properties of forests [Naesset et al., 2005], and retrieving leaf fuel moisture [Li et al., 2007]. PLS is a very popular method because it can cope with the existing high correlation between variables in a very efficient way.

The limitations of these regression tools have been circumvented by the introduction of nonlinear methods. Neural networks have been mainly used in combination of RTMs (we will come back to this hybrid approaches later). Nonlinear extension of PLS has been introduced via kernels in [Arenas-García and Camps-Valls, 2008]. Recently, the support vector regression (SVR) [Smola and Schölkopf, 2004] has yielded good results in modeling many biophysical parameters. In land applications, SVR is becoming a standard method that has proven high efficiency in modelling LAI [Durbha et al., 2007]. Several novel formulations of SVR have been recently presented, ranging from semisupervised methods (see Section 4.5.1) that exploit the wealth of unlabeled data in addition to the few labeled points [Bazi and Melgani, 2007, Camps-Valls et al., 2009], to multiouput SVR regression to provide estimations of several variables of interest simultaneously [Tuia et al., 2011b]. A review of SVR in remote sensing can be found in [Mountrakis et al., 2011].

In the recent years, the introduction of Gaussian Processes (GPs) [Rasmussen and Williams, 2006] techniques has alleviated some shortcomings of the previous methods. For instance, training GPs is far more simpler than neural networks. In addition, unlike in SVR, one can optimize very flexible kernels by maximum likelihood estimation. GPs generally yield very good numerical performance and stability. An application of GP to the estimation of land parameters can be found in [Furfaro et al., 2006]. Besides, GPs may be of particular interest because they do not only provide pointwise predictions but also confidence intervals for the predictions. A GP with a kernel adapted to both signal and noise properties was proposed for land parameter estimation in [Verrelst et al., 2011]. The method can provide a ranking of features (bands) and samples (spectra) for the task at hand. Section 6.5 gives empirical evidence of the GP advantages compared to other parametric and nonparametric methods.

6.2.2 OCEAN INVERSION MODELS

Open waters cover about three quarters of our planet's surface [Campbell, 2007]. Monitoring the processes involved in oceans, rivers, lakes or ice and snow covers is of paramount relevance for both managing human settlements and controlling planet changes. In this context, the field is vast and many model inversion methods have been developed, mainly related to specific sensors or particular applications. This section briefly reviews the main contributions in hydrospheric sciences according to a series of application areas.

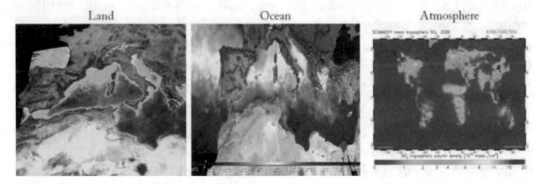

Figure 6.3: Sample products illustrating land, ocean and atmospheric products obtained by model inversion from satellite sensors. *Left:* AVHRR vegetation index over land and CZCS Ocean Colour phytoplankton concentration over sea in Europe and North Africa. *Middle:* Mediterranean sea-surface temperature from the Medspiration products. ESA's Sentinel-3 mission for GMES will also provide data on sea-surface temperature. *Right:* Mean nitrogen dioxide (NO_2) pollution map for 2006,

measured by Envisat's Scanning Imaging Absorption Spectrometer for Atmospheric Chartography (SCIAMACHY) instrument, which records the spectrum of sunlight shining through the atmosphere. Source: ESA, `http://www.esa.int`

Parametric methods

In oceanic models, the particular characteristics of color or ocean salinity can be of interest [O'Reilly et al., 1998]. Between the 1970 and the 1990, Landsat images were introduced to inventory, locate and identify water. In Philipson and Hafker [1981], Landsat MSS images were used to identify flooded areas: a very simple ratio comparing the brightness in the near infrared (band 4) was implemented. Later, in [Frazier and Page, 2000], also included band 5 to identify turbid water and separate such areas from the unflooded land. It should be noted that several researchers have focused on detecting coastal or riverine floods (e.g., see `http://floodobservatory.colorado.edu/`). Another interesting issue is that of finding water in desert areas. Geologists have used remote sensing images to identify the new supplies of water in the rocks beneath the Sahara desert[1].

The sea surface salinity (SSS) parameter is one of the most relevant ones in ocean remote sensing [Lagerloef et al., 1995]. The flow intrusions with low salinity influence the deep thermohaline circulation, the meridional heat transport, and the near-surface dynamics of tropical oceans. Also, it has been argued that salinity is related to El Niño dynamics. The problems and limitations identified 15 years ago are still present. Nevertheless, the recent ESAâL™s SMOS Mission [Kerr et al., 2001] has been launched to demonstrate the use of interferometry to observe salinity over oceans and ice characteristics, among other parameters. The main retrieval method uses a ratio-based parametrization of the measured brightness temperature of the Earth's surface at horizontal and vertical polarizations [Font et al., 2004].

The estimation of the sea surface temperature (SST) plays a crucial role in many environmental models, such as those estimating energy and mass exchanges between atmosphere and surface, numerical weather prediction models or climatic variability and global ocean circulation models [Dash et al., 2002]. Many algorithms have been proposed for retrieving SST from space radiometry [Haines et al., 2007]. Most of them are derived from radiative transfer theory and require prior information about the state of the atmosphere (e.g., temperature and water vapor profiles) [Barton, 1992, Chan and Gao, 2005]. Figure 6.3[middle] shows an illustrative example of a sea-surface temperature map obtained from the Medspiration products[2].

Empirical ocean color algorithms for deriving chlorophyll concentrations have been developed primarily using spectral blue-to-green ratios and have been validated with field data collected from a number of stations. Garver and Siegel [1997] developed a nonlinear statistical method for the inversion of ocean color data, which assumed the known spectral shapes of specific absorption coefficients. The model was later improved in [Maritorena and O'Reilly, 2000, O'Reilly et al., 1998]. Lee et al. [2002] developed a multiband quasi-analytical model involving low computational cost and moderate accuracy. Recently, Shanmugam et al. [2011] introduced a new inversion model to retrieve the particulate backscattering in both coastal and ocean waters. The method is compared to a quasi-analytical algorithm (QAA), a constrained linear matrix inversion model (LM) with statistical selection, and the Garver-Siegel-Maritorena's model. In Section 6.5, we also consider the problem of retrieval of oceanic chlorophyll concentrations.

Water quality assessment is another interesting application field for remote sensing parameter retrieval. In [Khorram, 1980], Landsat multispectral scanner (MSS) data, and U-2 colour and colour infrared photographs were combined with *in-situ* data for the assessment of water quality parameters: the water quality parameters of interest included turbidity and suspended solids. Later, in [Zhang and Chen, 2002], authors used a semi-empirical algorithm of water transparency with concurrent *in situ* measurements, Landsat TM and simulated SeaWiFS.

Finally, we should also mention the related fields of bathymetry and altimetry, where parametric models are extensively used. On the one hand, the field of bathymetry is concerned with the description of the water depth and ocean's floor. Landsat MSS1 bands 4 and 5 have been used to define depth-finding algorithms [Polcyn and Lyzenga, 1979]. Other parametric models commonly used involve other MSS bands [Jupp et al., 1985].

Lyzenga et al. [1979] suggested logarithmic relations between depth zones and reflectance values, while polynomial regression was used in [Polcyn and Lyzenga, 1979]. On the other hand, in the field of altimetry, one measures sea surface slopes and currents. The main parameters involved in the model inversion are wave height, wind speed and sea levels. Most of the parameters have a seasonal behavior so, in the recent years, multitemporal images are jointly exploited to improve estimations. An interesting application of altimetry along with atmosphere modeling is the observation of hurricanes since both temperature and sea height are altered by its presence.

Non-parametric methods

Non-parametric and non-linear regression techniques have been effectively introduced for the estimation of ocean biophysical parameters. For example, neural networks have been considered for the estimation of chlorophyll concentration [Bruzzone and Melgani, 2005, Keiner, 1999], water quality [Cipollini et al., 2001] or ocean color in coastal waters [Camps-Valls and Bruzzone, 2009, Dzwonkowski and Yan, 2005]. SVR have been applied successfully for coastal chlorophyll concentration in [Bruzzone and Melgani, 2005, Camps-Valls et al., 2006a, Kwiatkowska and Fargion, 2003, Zhan, 2005]. Non-parametric Bayesian approaches have also been used for retrieval. The standard formulation of the Relevance Vector Machine (RVM) [Tipping, 2001] was modified in [Camps-Valls et al., 2006c] to incorporate prior knowledge in the iterative optimization procedure, and showed good performance for the estimation of ocean chlorophyll concentration. Finally, GP have also been considered for this task in [Pasolli et al., 2010].

6.2.3 ATMOSPHERE INVERSION MODELS

Atmospheric modeling involves estimation of pollution indicators, trace gases (such as ozone or metane), temperature and water vapor [Rodgers, 2000].

Parametric methods

With the introduction of meteorological satellites, the possibility to estimate temperature and moisture profiles became possible. Kapla [1959] studied parametric relations between the vertical distribution of temperature in the atmosphere and the measured thermal emission from the 15μm band of CO_2. Also, King [1956] estimated the temperature profile by measuring the angular distribution of emitted radiation in a single spectral interval. A detailed method for atmospheric remote sensing sounding from satellites appeared in [Wark, 1961, Yamamoto, 1961]. More recently, thanks to the launching of the first sounders, such as the Infrared Interferometer Spectrometer (IRIS) or the Satellite Infrared Spectrometer (SIRS), the field of model inversion developed quite rapidly. Current sounders like AIRS or IASI have demanded more studies in model inversion due to the challenges posed by the huge data volume acquired, which makes the ill-posed problem even harder.

There are many parametric methods available in the literature used for doing model inversion. The Bayesian approach to estimation basically involves including prior knowledge about the state vector $\mathbf{X}$ (e.g., the temperature profile) given the radiances $\mathbf{y}$. By applying Bayes' rule, a reasonable selection for the estimated state vector $\hat{\mathbf{x}}$ is the value of $P(\mathbf{X}|\mathbf{y})$ that has the highest probability, i.e., the maximum a posteriori (MAP) estimator. An alternative to MAP consists of the minimization of a loss function. The MSE is commonly adopted and results in the called Bayes Least Squares (BLS) [Blackwell, 2002]. This approach is inefficient because it requires assuming (or having access to) a complete representation of the relation between $\mathbf{y}$ and $\mathbf{X}$, which is not typically available. This is why sometimes one constrains the estimator to be linear, thus leading to the linear least squares estimator (LS). However, in that case, only second order relations are taken into account.

The previous methods make use only of the statistical relations between state vectors and radiances. Other parametric approaches exploit the forward model as well. One example is the minimum information retrieval (MIR), which only depends on a nominal state $\mathbf{X}_o$. An alternative to this method is the maximum-resolution retrieval (MRR), which assumes a forward model and a linear retrieval model. The method can inherit the classical problem of maximizing the resolution while simultaneously minimizing the noise [Blackwell, 2002]. This is why some regularized and constrained versions of MRR exist in the literature.

In the last decade, researchers have turned to the use of nonlinear methods such as neural networks or support vector machines for retrieval. These methods typically overcome many of the aforementioned problems. Figure 6.3[right] shows an illustrative example of the estimated mean map of nitrogen dioxide (NO_2) pollution from Envisat's SCIAMACHY instrument[3].

Non-parametric methods

Neural networks have been considered for the estimation of temperature and water vapor profiles using hyperspectral sounding data [Aires, 2002, Blackwell, 2005]. The classical multilayer perceptron was chosen in both cases. In [Camps-Valls et al., 2011], results of standard multilayer neural networks, cascade neural nets and least squares SVM are given for the estimation of temperature and moisture profiles (see Section 6.5). NO_2 emissions from biomass burning was estimated in [Bruzzone et al., 2003], while the challenging problem of hourly surface ozone concentrations was successfully addressed with different topologies of dynamic neural networks in [Balaguer-Ballester et al., 2002]. SVR have also been considered for extracting atmospheric properties [Camps-Valls et al., 2011] or continental evapotranspiration processes [Yang et al., 2006]. In [Xie et al., 2008], SVR was used to estimate horizontal water vapor transport from multisensor data.

6.3 PHYSICAL INVERSION TECHNIQUES

The previous statistical approaches may lack transferability, generality, or robustness to new geographical areas. This has led to the advancement of physical models for estimating bio-geo-chemical structural state variables from spectra. The advantage of physical models is that they can be coupled from lower to higher levels (e.g., canopy level models build upon leaf models), thereby providing a physically-based linkage between optical EO data and biochemical or structural state variables [Jacquemoud et al., 2009, Verhoef and Bach, 2003]. RTMs can be applied in many ways for forestry applications; some can be run in the forward mode, while others in both forward and inverse modes:

- Running RT models in its direct or forward mode enables creating a database covering a wide range of situations and configurations. Such forward RT model simulations allow for sensitivity studies of canopy parameters relative to diverse observation specifications, which can lead to an improved understanding of the Earth Observation (EO) signal as well as to an optimized instrument design of future EO systems [Jacquemoud et al., 2000, Myneni et al., 1995].

- Inversion of the RT model enables retrieving particular characteristics from EO data. The unique and explicit solution for a model inversion depends on the number of free model parameters relative to the number of available independent observations. A prerequisite for a successful inversion is therefore the choice of a validated and appropriate RTM, which correctly represents the radiative transfer within the observed target [Myneni et al., 1995]. When a unique solution is not achieved then more *a priori* information is required to overcome the ill-posed problem.

The basic assumption for inverting RTMs is that the forward model contains all the necessary information about the problem, so its inversion may lead to accurate parameter estimates. There exist many RTMs implemented in software packages to deal with the forward modeling. For example, PROSPECT is an RTM accounting for leaf optical properties [Jacquemoud and Baret, 1990] while SAIL constitutes a canopy bidirectional reflectance model [Verhoef, 1984]. Their combination led to the PROSAIL model: advances on joint strategies are revised in [Jacquemoud et al., 2009]. While PROSPECT and SAIL are RTMs more focused on land/vegetation quantitative problems, other RTMs are more specific to ocean or atmosphere applications. For example, the optimal spectral sampling (OSS) is a fast and accurate monochromatic RTM well-suited for geophysical parameter retrieval [Moncet et al., 2008].

The design of accurate forward models is subject of active research [Asrar, 1989, Goel, 1989, Kimes et al., 2000, Liang, 2004, Pinty and

Verstraete, 1991]. In this chapter, however, rather than reviewing the physical aspects of RTMs, we will focus on the different families of inversion methodologies, their assumptions, advantages and shortcomings, along with the common applications. Several inversion approaches have been considered in the literature [Baret and Buis, 2008, Liang, 2004, 2008]: classical iterative optimization, look-up-tables (LUTs), simulated annealing and genetic algorithms [Fang and Liang, 2003, Renders and Flasse, 1996], Markov chain Monte Carlo (MCMC) [Zhang et al., 2005b], and alternative Bayesian approaches. Nevertheless, iterative optimization and (or over) LUTs are the most widely used [Liang, 2008].

6.3.1 OPTIMIZATION INVERSION METHODS

Optimization methods tackle the inverse problem in Eq. (6.1) by minimizing a loss function, $\mathcal{L}(\theta)$, respect to the state parameters, $\boldsymbol{\theta}$. To do this, a set of n data pairs generated by a particular RTM, $\{\mathbf{y}, \mathbf{X}\}_{i=1}^{n}$, are used. Different norms of the residuals can be used here. Some authors assume a Gaussian model for the uncertainties (residuals), thus minimizing the ℓ_2-norm, $\min_{\theta}\{\|\mathbf{y} - f(\mathbf{X}; \theta)\|^2\}$. Nevertheless, the most important issue is the inclusion of *prior* knowledge in the minimization functional accounting for model uncertainties, and several norms have been used in this sense [Bacour et al., 2002]. Other authors impose sparsity constraints or the (reasonable) positivity of the retrievals. The goal of these approaches is not only to obtain more realistic estimates but improving the convexity of the error surface [Combal, 2002].

In most of the cases, however, the most accurate physically-based models are non-differentiable and involve computationally expensive inversion methods. Classical techniques include: *simplex methods*, which are particularly useful for discontinuous, noisy or numerically-deficient functions; the (iterative) *conjugate direction set* method [Press et al., 1989], which may be suitable when many parameters are involved in the inversion; or *adjoint models* that provide (approximate) analytical expressions for the loss gradient (see [Liang, 2008] and references therein). In any case, the main problems of optimization inversion methods are typically related to the high computational cost involved, and the risk of falling into local minima. While the latter has been recently alleviated [Bonnans et al., 2006], the former constitutes a serious drawback for the adoption in operational regional or global studies.

Despite all these disadvantages, these methods are widely used in gneral. A good review on optimization methods used in remote sensing for land applications can be found in [Bacour et al., 2002]. Their main advantage is the flexibility and the easy inclusion of prior knowledge, while the main applications involve estimating vegetation parameters. For example, numerical inversion on POLarization and Directionality of the Earth's Reflectances (POLDER)[4] data was presented in Bicheron and Leroy [1999] for retrieving both LAI and fraction of photosynthetically active radiation (FPAR), while several canopy reflectance models were compared in [Jacquemoud et al., 2000] to estimate chlorophyll content, the leaf area index, and the mean leaf inclination angle. Later, in [Zarco-Tejada et al., 2001], hyperspectral data was used for chlorophyll content estimation using a numerical model inversion approach coupled to the PROSPECT model. Houborg et al. [2007] retrieved LAI, chlorophyll content, and total vegetation water content using Terra and Aqua MODIS multi-spectral, multi-temporal, and multi-angle reflectance observations. In Gemmell et al. [2002], boreal forest cover and tree crown transparency was obtained from Landsat TM data. In this case, inclusion of ancillary data became necessary to regularize the solution.

6.3.2 GENETIC ALGORITHMS

The main problems faced in the previous optimization approaches are the convergence to an optimal solution, the dependence on the initialization parameters of the algorithms, and the associated high computational cost. Genetic algorithms [Goldberg, 1989] can be an alternative solution. Unfortunately, very few contributions are found in the literature so far: retrieval of land surface roughness and soil moisture [Wang and Jin, 2000], forest parametrization [Lin and Sarabandi, 1999], LAI estimation from either optical [Fang and Liang, 2003] or thermal infrared multiangle data [Zhuang and Xu, 2000]. In [Lin and Sarabandi, 1999], GA are used for the retrieval of forest parameters using a fractal-based coherent scattering model. Also water optical properties have been derived [Slade et al., 2004, Zhan et al., 2003].

6.3.3 LOOK-UP TABLES

This approach considers constructing a database, or Look-up table (LUT), of representative spectra and associated state parameters. For every new incoming spectra, a simple criterion of search can be applied to assign the most appropriate parameter. The advantages are clear: building the LUT is an off-line process so computationally cheap, the minimization criterion (typically based on RMSE) is simple to apply and leads to a global solution. Behind this theoretical simplicity, important problems arise [Weiss et al., 2000]. First, building a representative LUT that accounts for most of uncertainties and properly covers the parameter space is a difficult task [Combal, 2002, Weiss et al., 2000]. These problems give rise to very big LUTs, that arc hardly manageable which, in turn, inducing slow searches. Second, the approach typically relies on heuristic criteria for both selecting the most representative cases (pairs of spectra and associated parameters) [Combal, 2002, Weiss et al., 2000], and the search criterion [Knyazikhin, 1998a, Knyazikhin et al., 1998].

Mainly due to its efficiency, the approach is widely adopted in prospective studies and in operational product generation, such as for retrieving LAI and FPAR products from surface reflectances in the Moderate Resolution Imaging Spectroradiometer (MODIS)[5] and the Multiangle Imaging SpectroRadiometer (MISR)[6] [Knyazikhin, 1999, Myneni et al., 2002]. Darvishzadeh et al. [2008] obtained LAI and leaf and canopy chlorophyll contents by inverting the PROSAIL model. This is a very popular approach, mainly for works addressing a single test site (e.g., with airborne data). In [González-Sanpedro et al., 2008], LAI was derived from a multitemporal series of images inverting the PROSPECT+SAIL model. In this case, the LUT was designed to take into account the different crop types present in the area.

6.3.4 BAYESIAN METHODS

Model inversion following Bayesian approaches try to approximate the posterior distribution, $p(\mathbf{X}|\mathbf{y})$, i.e., the distribution of the variables given the reflectance/radiance. The main advantages rely on the fact that no assumption about data distributions need to be done. Nevertheless, the approach has been scarcely adopted, mainly due to high computational cost of iterative approaches such as the Markov chain Monte Carlo (MCMC) [Zhang et al., 2005b]. The alternative Importance Sampling (IS) [Makowski et al., 2006] alleviates this problem but has not been extensively evaluated in remote sensing applications. A Bayesian estimation for land surface temperature retrieval was presented in [Morgan, 2005], and for soil moisture retrieval in [Notarnicola et al., 2008].

6.4 HYBRID INVERSION METHODS

In recent years, combined approaches that use the databases generated by RTMs and powerful regression approaches have been introduced in the literature. These *hybrid methods* try to exploit the advantages of physically-based models and the flexibility and computational efficiency of nonparametric nonlinear regression methods. The idea is to learn the inverse mapping g of Eq. (6.2) with, for example, neural networks using data generated by accurate RTMs. Several nonparametric methods have been applied to invert RTMs, from decision trees to neural networks and kernel methods. The main issue is how representative is the dataset because inversion will not be possible (or deficient) if the parameter space is poorly sampled. This poses several related problems: how many data points are necessary for representing the problem, how many input (parameters) and output (spectra) dimensions should enter the RTM, and how should one train regression methods in such potentially complex situations. Several families of methods have been used so far, which are briefly summarized below.

6.4.1 REGRESSION TREES

Regression trees have been used to estimate land surface variables: Weiss and Baret [19991] retrieved LAI, fraction of photo-synthetically active radiation (FAPAR) and chlorophyll content from VEGETATION/SPOT4[7];

Gong et al. [1999] inverted LAI from a simulated database; Fang et al. [2003] inverted LAI from the Landsat ETM imagery[8]; Liang et al. [2003] estimated LAI and broadband albedo from the Earth Observing 1 (EO-1)[9] data by incorporating the soil line concept; Kimes et al. [2002] inverted a complex 3D DART model for a wide range of simulated forest canopies using POLDER-like data; and broadband land surface albedo was estimated in [Liang et al., 2003].

6.4.2 NEURAL NETWORKS

The vast majority of hybrid inversion methods consider the use of neural networks [Baret and Fourty, 1997, Baret et al., 1995, Gopal and Woodcock, 1996, Jin and Liu, 1997, Kimes et al., 1998, Smith, 1993] for retrieval of canopy parameters. Actually, with the current neural network implementations, they are fast to train a test. Vasudevan et al. [2004] used neural nets for deriving integrated water vapor and cloud liquid water contents over oceans from brightness temperatures measured by the Multifrequency Scanning Microwave Radiometer (MSMR). Fang and Liang [2005] noted deficiencies in LAI retrieval by standard LUT methods, and compared neural nets and projection pursuit to obtain less biased estimates. In [Bacour et al., 2006], the network was trained on a reflectance database made of RTM simulations, and LAI, FAPAR and FCOVER were accurately retrieved as compared with ground measurements. Very recently, the combination of clustering and neural networks inverted simulated data with additive noise. Inclusion of multiangle images improved the LAI estimations. Lately, in [Verger et al., 2011], neural networks were successfully developed over RTMs to estimate LAI, FCOVER and FAPAR.

6.4.3 KERNEL METHODS

Only very recently, the use of kernel methods [Camps-Valls and Bruzzone, 2009], such as the support vector regression (SVR) or the kernel ridge regression (KRR) have been used. In [Durbha et al., 2007], LAI is retrieved by inverting PROSAIL using SVR over Multi-angle Imaging SpectroRadiometer (MISR) data[10]. In our recent work [Camps-Valls et al.,

2011], linear regression, multilayer perceptrons, cascade neural nets and kernel ridge regression were used to invert the OSS transfer model for atmospheric parameter retrieval. The advantage of the approach allowed improved accuracy versus standard *optimal estimation* approaches[11], and great computational savings for inversion and retrieval.

6.5 EXPERIMENTS

This section illustrates the application of the three inversion approaches revised (statistical, physical and hybrid) in land, ocean and atmospheric parameter retrieval applications.

6.5.1 LAND SURFACE BIOPHYSICAL PARAMETER ESTIMATION

The first application deals with the *statistical approach* to retrieve vegetation parameters using *in situ* measurements and hyperspectral images. We compare the most representative parametric approaches, and state-of-the-art nonparametric methods, such as SVR [Smola and Schölkopf, 2004], GP [Rasmussen and Williams, 2006, Verrelst et al., 2011] and the recently presented multi-output SVR (MSVR) [Tuia et al., 2011b].

The data were obtained in the SPARC-2003 (SPectra bARrax Campaign) and SPARC-2004 campaigns in Barrax, La Mancha, Spain. The region consists of approximately 65% dry land and 35% irrigated land. The methodology applied to obtain the *in situ* leaf-level Chl_{ab} data consisted of measuring samples with a calibrated CCM-200 Chlorophyll Content Meter in the field. Concurrently, we used CHRIS images Mode 1 (62 spectral bands, 34m spatial resolution at nadir). The images were preprocessed [Gómez-Chova et al., 2008], geometrically [Alonso and Moreno, 2005] and atmospherically corrected [Guanter et al., 2005]. Summarizing, a total of n = 136 datapoints in a 62-dimensional space and the measured chlorophyll concentration constitute the database.

Performances of a wide array of established indices, linear regression with all bands, SVR, MSVR and GP were tested. Models were run for a total of 50 random realizations of the training and test data. Averaged correlation coefficients are shown for the test set in Table 6.1. It is clearly observed that nonparametric methods show the best results in both correlation and stability, with GP performing best of the tested methods.

Table 6.1: Correlation coefficient R results of narrowband and broadband indices proposed in relevant literature tested in the present study along with recent nonparametric models. See [Verrelst et al., 2011] and references therein.

Method	Formulation	R
GI	R_{672}/R_{550}	0.52 (0.09)
GVI	$(R_{682}-R_{553})/(R_{682}+R_{553})$	0.66 (0.07)
Macc	$(R_{780}-R_{710})/(R_{780}+R_{680})$	0.20 (0.29)
MCARI	$[(R_{700}-R_{670})-0.2(R_{700}-R_{550})]/(R_{700}/R_{670})$	0.35 (0.14)
MCARI2	$1.2[2.5(R_{800}-R_{670})-1.3(R_{800}-R_{550})]$	0.71 (0.12)
mNDVI	$(R_{800}-R_{680})/(R_{800}+R_{680}-2R_{445})$	0.77 (0.12)
mNDVI$_{705}$	$(R_{750}-R_{705})/(R_{750}+R_{705}-2R_{445})$	0.80 (0.07)
mSR$_{705}$	$(R_{750}-R_{445})/(R_{705}+R_{445})$	0.72 (0.07)
MTCI	$(R_{754}-R_{709})/(R_{709}+R_{681})$	0.19 (0.26)
mTVI	$1.2[1.2(R_{800}-R_{550})-2.5(R_{670}-R_{550})])$	0.73 (0.07)
NDVI	$(R_{800}-R_{670})/(R_{800}+R_{670})$	0.77 (0.08)
NDVI2	$(R_{750}-R_{705})/(R_{750}+R_{705})$	0.81 (0.06)
NPCI	$(R_{680}-R_{430})/(R_{680}+R_{430})$	0.72 (0.08)
NPQI	$(R_{415}-R_{435})/(R_{415}+R_{435})$	0.61 (0.15)
OSAVI	$1.16(R_{800}-R_{670})/(R_{800}+R_{670}+0.16)$	0.79 (0.09)
PRI	$(R_{531}-R_{570})/(R_{531}+R_{570})$	0.77 (0.07)
PRI2	$(R_{570}-R_{539})/(R_{570}+R_{539})$	0.76 (0.07)
PSRI	$(R_{680}-R_{500})/R_{750}$	0.79 (0.08)
RDVI	$(R_{800}-R_{670})/\sqrt{(R_{800}+R_{670})}$	0.76 (0.08)
SIPI	$(R_{800}-R_{445})/(R_{800}-R_{680})$	0.78 (0.08)
SPVI	$0.4[3.7(R_{800}-R_{670})-1.2(R_{530}-R_{670})]$	0.70 (0.08)
SR	R_{800}/R_{680}	0.63 (0.12)
SR1	R_{750}/R_{700}	0.74 (0.07)
SR2	R_{752}/R_{690}	0.68 (0.09)
SR3	R_{750}/R_{550}	0.75 (0.07)
SR4	R_{672}/R_{550}	0.76 (0.10)
SRPI	R_{430}/R_{680}	0.76 (0.09)
TCARI	$3[R_{700}-R_{670})-0.2(R_{700}-R_{550})(R_{700}/R_{670})]$	0.53 (0.13)
TVI	$0.5[120(R_{750}-R_{550})-200(R_{670}-R_{550})]$	0.70 (0.10)
VOG	$R_{740}/(R_{720}$	0.76 (0.06)
VOG2	$(R_{734}-R_{747})/(R_{715}+R_{726})$	0.72 (0.09)
NAOC	Area in [643, 795]	0.79 (0.09)
LR	Least squares	0.88 (0.06)
SVR [Smola and Schölkopf, 2004]	RBF kernel	0.98 (0.03)
MSVR [Tuia et al., 2011b]	RBF kernel	0.98 (0.03)
GP [Verrelst et al., 2011]	Anisotropic RBF kernel	0.99 (0.02)

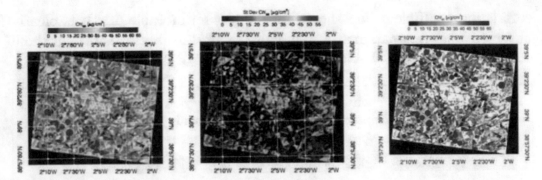

Figure 6.4: Final chlorophyll concentration prediction map (left) and predictive standard deviation (middle) and masked confidence map (right) generated with GP on the CHRIS 12-07-2003 nadir image.

The best GP model was used for prediction on the whole CHRIS image to generate a pixel-by-pixel map of *Chl* and its confidence map (see Fig. 6.4). The maps show clearly the irrigated crops (the circles in orange-red), the semi-natural areas (light blue) and the bare soil areas (dark blue). Although these maps cannot be used as a validation *per se*, the confidence maps allow us to draw conclusions on the performance of the retrievals. For example, the high confidences (western part of the image) were the fields sampled the most, while low confidence predictions (center of the image) correspond to areas particularly underrepresented in the training data, such as dry barley, harvested barley, and bright bare soils. This product may be used to set sensitivity margins of field instruments quite intuitively: areas are thresholded with error levels above 10% of the total *Chl* range (e.g., $>6.5 \mu g/cm^{-2}$), see Fig. 6.4[right]. These results confirm the very good properties of nonparametric models in general and GPs in particular for the estimation of land cover properties from space, applicable for both multispectral and hyperspectral sensors.

6.5.2 OPTICAL OCEANIC PARAMETER ESTIMATION

In this section, performance of neural networks, SVR, RVM and GP in the estimation of oceanic chlorophyll concentration from measured reflectances is compared in terms of accuracy, bias, and sparsity. The SeaBAM dataset is used [O'Reilly et al., 1998], which gathers 919 *in-situ* measurements of chlorophyll concentration around the United States and Europe. The dataset contains *in situ* pigments and remote sensing reflectance measurements at wavelengths present in the SeaWiFS sensor.[12]

Table 6.2 presents results in the test set for all models. For comparison purposes, we include results obtained with a feedforward neural network trained with back-propagation (NN-BP), which is a standard approach in biophysical parameters retrieval. Also, we include results for standard parametric models (Morel, CalCOFI and OC families) [Maritorena and O'Reilly, 2000]. We can observe that (1) SVR, RVM and GP show a better performance than empirical Morel and OC2 models, and also better than artificial neural networks (NN-BP); (2) the SVR and GP techniques are more accurate (RMSE, MAE); (3) RVM and GP are less biased (ME) than the rest of the models; and (4) in the case of the RVMs, drastically much more sparse (only 4.9% of training samples were necessary to attain good generalization capabilities). Comparing SVR and RVM, we can state that RVMs provide accurate estimations (similar to SVR) with small number of relevant vectors. Nevertheless, GPs provide more accurate results than SVR and RVM.

Table 6.2: Mean error (ME), root mean-squared error (RMSE), mean absolute error (MAE), and correlation coefficient between the actual and the estimated Chl-a concentration (r) of models in the test set. More results can be found in [Camps-Valls et al., 2006a, c]

Methods	ME	RMSE	MAE	r	[%]SVs/RVs
Morel-1[†,‡]	-0.023	0.178	0.139	0.956	–
Morel-3	-0.025	0.182	0.143	0.954	–
CalCOFI 2-band cubic	-0.051	0.177	0.142	0.960	–
CalCOFI 2-band linear	0.079	0.325	0.256	0.956	–
Ocean Chlorophyll 2, OC2	-0.031	0.169	0.133	0.960	–
Ocean Chlorophyll 4, OC4	-0.039	0.162	0.129	0.966	–
NN-BP, 4 hidden nodes	-0.046	0.143	0.111	**0.971**	–
SVR	-0.070	0.139	**0.105**	**0.971**	44.3%
RVM	**-0.009**	0.146	0.107	0.970	4.9%
GP	**-0.009**	**0.103**	0.107	0.961	–

[†] Morel-1 and CalCOFI 2-band linear are described by $C = 10^{a_0 + a_1 R}$, Morel-3 and CalCOFI 2-band cubic are cubic interpolation described by $C = 10^{a_0 + a_1 R + a_2 R^2 + a_3 R^3}$, and models OC2/OC4 are described by $C = a_0 + 10^{a_1 + a_2 R + a_3 R^2 + a_4 R^3}$, where for Morel models $R = \log(Rrs443/Rrs555)$, for CalCOFI and OC2 models $R = \log(Rrs490/Rrs555)$, and for OC4 model $R = \log(\max[Rrs443, Rrs490, Rrs510]/Rrs555)$.

[‡] The results provided in this table for Morel, CalCOFI and OC2/OC4 models slightly differ from the ones given in [O'Reilly et al., 1998] since they are computed only for the test samples. In addition, models in [O'Reilly et al., 1998] used all available data to fit the models and hence no validation procedure was followed.

These results suggest that GPs are potentially useful methods for ocean chlorophyll concentration estimation. We did a one-way analysis of variance (ANOVA) to compare the means of the models and no significant statistical differences were observed between the nonparametric approaches for accuracy ($p = 0.31$) and bias ($p = 0.33$). Therefore, as a preliminary conclusion, we can state that the nonparametric solutions are more accurate than parametric or physical models.

Figure 6.5 shows a comparison between the predictions obtained by the GP model and the OC4 model. In the same figure, we also show the predicted values against the residuals for these models. These goodness-of-fit graphics confirm the good predictive performance of both models. However, some differences can be observed. Excellent coefficient of determination ($R^2 = 0.955$, test set) and unbiased estimations (Slope: 1.001 ± 0.010, intercept: $0.001\pm3e-9$) and errors are observed for the RVM, which are slightly better than those shown by the OC4 model ($R^2 = 0.933$, Slope: 1.088 ± 0.034, intercept: $0.055\pm3e-5$). In particular, it is worth to note the much higher bias in the prediction errors versus residuals obtained by the OC4 (compare fitted regression lines in Fig. 6.5(b) and (d)). This is an important second characteristic of the model since unbiased estimations are desirable in estimating biophysical parameter, see the mean error in Table 6.2.

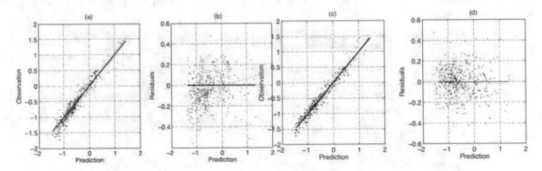

Figure 6.5: Performance of the OC4 (a,b) and GP (c,d) in the test set. Predicted versus observed concentrations, and predicted versus residuals are depicted.

6.5.3 MODEL INVERSION OF ATMOSPHERIC SOUNDING DATA

Temperature and water vapor are atmospheric parameters of high importance for weather forecast and atmospheric chemistry studies [Hilton et al., 2009, Liou, 2002]. Nowadays, observations from spaceborne high spectral resolution infrared sounding instruments can be used to calculate the profiles of such atmospheric parameters with unprecedented accuracy and vertical resolution [Huang et al., 1992]. In this section we use data acquired by the Infrared Atmospheric Sounding Interferometer (IASI) [Siméoni et al., 1997] (see Chapter 2 for details).

This section presents results on the use of linear regression (LR), multilayer neural networks (NN), cascade neural networks (CNN), and kernel ridge regression (KRR) to derive temperature and moisture profiles from hyperspectral infrared sounding spectra [Camps-Valls et al., 2011]. We followed an hybrid approach: first we generated data using an RTM and then trained the nonparametric regression models. Test results are shown on a full IASI orbit (91800 IFOVs in 2008-03-04). As ancillary data we used AVHRR data, which has a much finer spatial resolution and thus allows proper visual inspection and comparison of the results. A total amount of 67475 synthetic FOVs were simulated by the European Organisation for the Exploitation of Meteorological Satellites (EUMETSAT)[13] with the OSS infrared radiative transfer code [Moncet et al., 2008] according to input profiles of temperature and humidity given in 90 pressure levels. The state vector (output space) to be retrieved consists of temperature and water vapor profiles. The range between [100, 1000] hPa is sampled at 90 pressure levels, thus constituting a challenging nonlinear multi-output regression problem.

A total of 4699 channels were retained by removing noisy channels [Calbet and Schlüssel, 2005]. Then, PCA was performed and the optimal number of components selected by training LR and KRR with standard cross-validation. The best results are shown in Fig. 6.6. The optimal linear combination of models in terms of MSE is also given. All nonlinear models outperform LR. Neural nets and KRR perform similarly, and, in general, no numerical differences (both in terms of RMSE and bias) are observed, while the OLC method reports an important gain for temperature estimation, yielding an average gain in RMSE of +0.6K. Model combination not only gives smoother error profiles (RMSE<1K) but also lower biases (ME<0.5K).

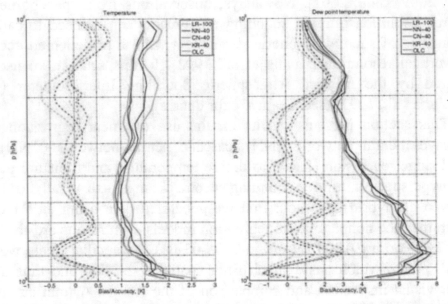

Figure 6.6: RMSE (solid) and bias (dashed) quality measures obtained with linear and nonlinear regression for the case of real MetOp-IASI data in cloud-free and emissivity 'sea' scenarios. The number of used features after PCA is indicated in the legend for each regression method.

Figure 6.7 illustrates the results obtained by KRR in some areas of interest. For the case of Madagascar, it is clear that big errors in temperature estimation, ΔT, highly correlate with the presence of clouds. Low confidence values are assigned to cloudy regions, see e.g., north-east Madagascar and south-west Mexico images.

6.6 SUMMARY

This chapter revised one of the most active fields in remote sensing image processing: the estimation of physical parameters from acquired images and *in-situ* measurements. We presented the main approaches in the literature and introduced the principles and standard terminology. A thorough comparison between all approaches is still missing in the literature but it seems that hybrid methods embracing physics and statistics provide accurate and computationally efficient methods. Several real application examples were provided in land, ocean and atmospheric parameter retrieval, and confirmed the excellent results obtained by nonparametric statistical approaches, either alone or in combination with radiative transfer models. The future poses some challenging problems: just to name a few, we will be confronted to the availability of huge amount of training data, to the design of more sophisticated and realistic RTMs, to the combination of both statistical and physically-based models, and to the specification of models that can adapt to multitemporal domains.

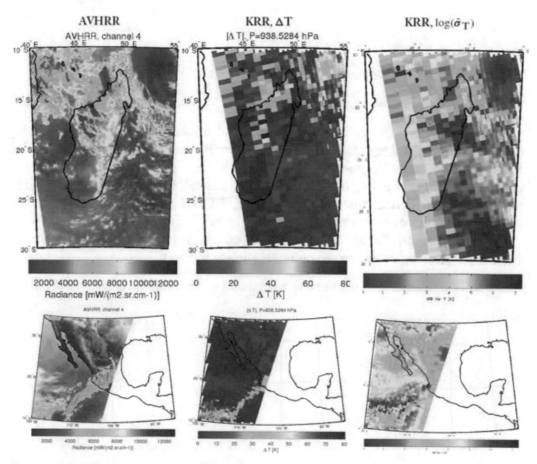

Figure 6.7: Illustrative examples of prediction errors for Madagascar [top] and Mexico [bottom]. We show the AVHRR radiance in channel 4 for visual inspection, the discrepancies of KRR predictions with the ECMWF ground truth, and the KRR predictive variance (in log-scale for proper visualization), which was computed as in GPs.

[1] http://www.nae.edu/File.aspx?id=7423

[2] http://projets.ifremer.fr/cersat/Information/Projects/
MEDSPIRATION2

[3] http://envisat.esa.int/instruments/sciamachy/

[4] http://smsc.cnes.fr/POLDER/

[5] http://modis.gsfc.nasa.gov/

[6] http://www-misr.jpl.nasa.gov/

[7] http://www.spot-vegetation.com/

[8] http://landsat.gsfc.nasa.gov/

[9] http://eol.gsfc.nasa.gov/

[10] http://www-misr.jpl.nasa.gov/

[11] Optimalestimation (OE) refers to a collection of inversion techniques for deriving physical properties from satellite measurements. The methods have relatively high accuracy and can incorporate prior knowledge. However, OE is a very computationally demanding method because, in general, the forward model has to be called iteratively several times for each pixel. Even for fast RTMs, like the OSS used here, this involves obtaining predictions in hours or even days. We address the reader to [Rodgers, 2000] for an excellent book on inverse methods for atmospheric sounding, including details on the optimal estimation theory.

[12] More information at http://seabass.gsfc.nasa.gov/seabam/seabam.html.

[13] www.eumetsat.int/

Bibliography

P.G. Abousleman, M.W. Marcellin, and R. B. Hunt. Compression of hyperspectral imagery using the 3-D DCT and hybrid DPCM/DCT. *IEEE Trans. Geosci. Rem. Sens.*, 33 (1): 26–34, 1995. DOI: 10.1109/36.368225 Cited on page(s) 29

J.B Adams, D.E Sabol, V Kapos, R.A Filho, D.A Roberts, M.O Smith, and A.R. Gillespie. Classification of multispectral images based on fractions of endmembers: Application to land-cover change in the Brazilian Amazon. *Rem. Sens. Environ.*, 52 (2): 137–154, 1995. DOI: 10.1016/0034-4257(94)00098-8 Cited on page(s) 84

J. Ahlberg and I. Renhorn. Multi- and hyperspectral target and anomaly detection. Technical Report FOI-R-1526-SE, Swedish Defence Research Agency, 2004. Cited on page(s) 71, 72

B. Aiazzi, L. Alparone, A. Barducci, S. Baronti, and I. Pippi. Estimating noise and information for multispectral imagery. *Optical Engineering*, 41: 656–668, March 2002. DOI: 10.1117/1.1447547 Cited on page(s) 21

F. Aires. A regularized neural net approach for retrieval of atmospheric and surface temperatures with the IASI instrument. *Journal of Applied Meteorology*, 41: 144–159, 2002. DOI: 10.1175/1520-0450(2002)041%3C0144:ARNNAF%3E2.0.CO;2 Cited on page(s) 111

H. Akaike. A new look at the statistical model identification. *IEEE Transactions Automat. Contr*, 19 (6): 716–723, 1974. DOI: 10.1109/TAC.1974.1100705 Cited on page(s) 89

L. Alonso and J. Moreno. Advances and limitations in a parametric geometric correction of CHRIS/PROBA data. In *Proceedings of the 3rd CHRIS/PROBA Workshop*, 2005. Cited on page(s) 116

L. Alparone, S. Baronti, A. Garzelli, and F. Nencini. A global quality measurement of pan-sharpened multispectral imagery. *IEEE Geosc. Rem. Sens. Lett.*, 1 (4): 313–317, oct. 2004. DOI: 10.1109/LGRS.2004.836784 Cited on page(s) 29

J. Amorós-López, L. Gómez-Chova, L. Alonso, J. Moreno, and G. Camps-Valls. Regularized multi-resolution spatial unmixing for MERIS and Landsat image fusion. *IEEE Geosc. Rem. Sens. Lett.*, 8 (1), 2011. DOI: 10.1109/IGARSS.2010.5649142 Cited on page(s) 86

R. Archibald and G. Fann. Feature selection and classification of hyperspectral images with support vector machines. *IEEE Geosc. Rem. Sens. Lett.*, 4 (4): 674–679, 2007. DOI: 10.1109/LGRS.2007.905116 Cited on page(s) 39

J. Arenas-García and G. Camps-Valls. Efficient kernel orthonormalized PLS for remote sensing applications. *IEEE Trans. Geosci. Rem. Sens.*, 46 (10): 2872–2881, Oct 2008. DOI: 10.1109/TGRS.2008.918765 Cited on page(s) 43, 107

J. Arenas-García and K.B. Petersen. *Kernel Methods for Remote Sensing Data Analysis*, chapter Kernel Multivariate Analysis in Remote Sensing Feature Extraction, pages 329–352. J. Wiley & Sons Inc., UK, 2009. Cited on page(s) 44

D.P. Argialas and C.A. Harlow. Computational image interpretation models: an overview and a perspective. *Photogrammetric Engineering & Remote Sensing*, 56 (6): 871–886, 1990. Cited on page(s) 50

G. Asrar. *Theory and Applications of Optical Remote Sensing*. John Wiley & Sons, Hoboken, NJ, 1989. Cited on page(s) 112

P.M Atkinson, M.E.J Cutler, and H. Lewis. Mapping sub-pixel proportional land cover with AVHRR imagery. *Int. Jour. Rem. Sens.*, 18 (4): 917–935, 1997. DOI: 10.1080/014311697218836 Cited on page(s) 97

M.F. Augusteijn, L.E. Clemens, and K.A. Shaw. Performance evaluation of texture measures for ground cover identification in satellite images by means of a neural network classifier. *IEEE Trans. Geosci. Rem. Sens.*, 33 (3): 616–626, May 1995. DOI: 10.1109/36.387577 Cited on page(s) 50

M. Awad, K. Chehdi, and A. Nasri. Multicomponent image segmentation using a genetic algorithm and artificial neural network. *IEEE Geosci. Remote Sens. Lett.*, 4 (4): 571–575, 2007. DOI: 10.1109/LGRS.2007.903064 Cited on page(s) 64

M. Baatz and Schäpe. Multiresolution segmentation: an optimization approach for high quality multi-scale image segmentation. In *Angewandte Geographische Informationsverarbeitung XII, Beiträge zum AGIT-Symposium*, pages 12–23. Herbert Wichmann Verlag, 2000. Cited on page(s) 65

C. Bachmann and T. Donato. An information theoretic comparison of projection pursuit and principal component features for classification of Landsat TM imagery of central Colorado. *Int. J. Rem. Sens*, 21 (15): 2927–2935(9), 2000. DOI: 10.1080/01431160050121339 Cited on page(s) 89

C.M. Bachmann, T.F. Donato, G.M. Lamela, W.J. Rhea, M.H. Bettenhausen, R.A. Fusina, K.R. Du Bois, J.H. Porter, and B.R. Truitt. Automatic classification of land cover on Smith Island, VA, using HyMAP imagery. *IEEE Trans. Geosci. Rem. Sens.*, 40 (10): 2313–2330, Oct 2002. DOI: 10.1109/TGRS.2002.804834 Cited on page(s) 64

C.M. Bachmann, T.L. Ainsworth, and R.A. Fusina. Exploiting manifold geometry in hyperspectral imagery. *IEEE Trans. Geosc. Rem. Sens.*, 43 (3): 441–454, Mar 2005. DOI: 10.1109/TGRS.2004.842292 Cited on page(s) 43, 89

C.M. Bachmann, T.L. Ainsworth, and R.A. Fusina. Improved manifold coordinate representations of large-scale hyperspectral scenes. *IEEE Trans. Geosc. Rem. Sens.*, 44 (10): 2786–2803, Oct 2006. DOI: 10.1109/TGRS.2006.881801 Cited on page(s) 43, 89

C. Bacour, S. Jacquemoud, M. Leroy, O. Hautecur, M. Weiss, L. Prévot, N. Bruguier, and H. Chauki. Reliability of the estimation of vegetation characteristics by inversion of three canopy reflectance models on airborne POLDER data. *Agronomie*, 22 (6): 555–565, 2002. DOI: 10.1051/agro:2002039 Cited on page(s) 112, 113

C. Bacour, F. Baret, D. Béal, M. Weiss, and K. Pavageau. Neural network estimation of LAI, fAPAR, fCover and LAI×Cab, from top of canopy MERIS reflectance data: Principles and validation. *Rem. Sens. Environ.*, 105 (4): 313–325, 2006. DOI: 10.1016/j.rse.2006.07.014 Cited on page(s) 115

E. Balaguer-Ballester, G. Camps-Valls, J. L. Carrasco-Rodríguez, E. Soria-Olivas, and S. del Valle-Tascon. Effective 1-day ahead prediction of hourly surface ozone concentrations in eastern Spain using linear models and neural networks. *Ecological Modelling*, 156: 27–41, 2002. DOI: 10.1016/S0304-3800(02)00127-8 Cited on page(s) 111

S. Bandyopadhyay, U. Maulik, and A. Mukhopadhyay. Multiobjective genetic clustering for pixel classification in remote sensing imagery. *IEEE Trans. Geosci. Rem. Sens.*, 45 (5): 1506–1511, 2007. DOI: 10.1109/TGRS.2007.892604 Cited on page(s) 64

A. Banerjee, P. Burlina, and C. Diehl. One-class SVM for hyperspectral anomaly detection. In G. Camps-Valls and L. Bruzzone, editors, *Kernel methods for remote sensing data analysis*, pages 169–192. J. Wiley & Sons, NJ, USA, 2009. Cited on page(s) 73

G. Baofeng, S.R. Gunn, R.I. Damper, and J.D.B. Nelson. Customizing kernel functions for SVM-based hyperspectral image classification. *IEEE Trans. Im. Proc.*, 17 (4): 622–629, April 2008. DOI: 10.1109/TIP.2008.918955 Cited on page(s) 64

A. Baraldi and F. Parmiggiani. An investigation of the textural characteristics associated with gray level cooccurrence matrix statistical parameters. *IEEE Trans. Geosci. Rem. Sens.*, 33 (2): 293–304, Mar 1995. DOI: 10.1109/36.377929 Cited on page(s) 51

A. Baraldi, V. Puzzolo, P. Blonda, L. Bruzzone, and C. Tarantino. Automatic spectral rule-based preliminary mapping of calibrated landsat TM and ETM+ images. *IEEE Trans. Geosci. Rem. Sens.*, 44 (9): 2563–2586, 2006. DOI: 10.1109/TGRS.2006.874140 Cited on page(s) 64

D.G. Barber and E.F. Ledrew. SAR sea ice discrimination using texture statistics: a multivariate approach. *Photogrammetric Engineering and Remote Sensing*, 57 (4): 385–395, 1991. Cited on page(s) 51

A. Barducci and I. Pippi. Analysis and rejection of systematic disturbances in hyperspectral remotely sensed images of the Earth. *Applied Optics*, 40 (9): 1464–1477, 2001. DOI: 10.1364/AO.40.001464 Cited on page(s) 21

F. Baret and S. Buis. Estimating canopy characteristics from remote sensing observations: Review of methods and associated problems. In *Advances in Land Remote Sensing: System, Modeling, Inversion and Applications*. Springer Verlag, Germany, 2008. DOI: 10.1007/978-1-4020-6450-0_7 Cited on page(s) 101, 102, 112

F. Baret and T Fourty. Estimation of leaf water content and specific leaf weight from reflectance and transmittance measurements. *Agronomie 17*, pages 455–464, 1997. DOI: 10.1051/agro:19970903 Cited on page(s) 115

F. Baret, J. Clevers, and M. Steven. The robustness of canopy gap fraction estimates from red and near-infrared retlectances: A comparison of approaches. *Rem. Sens. Environ. 54*, pages 141–151, 1995. DOI: 10.1016/0034-4257(95)00136-O Cited on page(s) 115

M.J. Barnsley, J.J. Settle, M. Cutter, D. Lobb, and F. Teston. The PROBA/CHRIS mission: a low-cost smallsat for hyperspectral, multi-angle, observations of the Earth surface and atmosphere. *IEEE Trans. Geosc. Rem. Sens.*, 42 (7): 1512–1520, Jul 2004. DOI: 10.1109/TGRS.2004.827260 Cited on page(s) 14

I.J. Barton. Satellite-derived sea surface temperature–A comparison between operational, theoretical, and experimental algorithms. *J. Applied Meteorology*, 31: 432–442, 1992. DOI: 10.1175/1520-0450(1992)031%3C0433:SDSSTA%3E2.0.CO;2 Cited on page(s) 109

C. A. Bateson and B. Curtiss. A Method for Manual Endmember Selection and Spectral Unmixing. *Rem. Sens. Env.*, 55 (3): 229–243, 1996. DOI: 10.1016/S0034-4257(95)00177-8 Cited on page(s) 92

C.A. Bateson, G.P. Asner, and C.A. Wessman. Endmember bundles: a new approach to incorporating endmember variability into spectral mixture analysis. *IEEE Trans. Geosc. Rem. Sens.*, 38 (2): 1083–1094, Mar 2000. DOI: 10.1109/36.841987 Cited on page(s) 93

T.C. Bau, S. Sarkar, and G. Healey. Hyperspectral region classification using a three-dimensional Gabor filterbank. *IEEE Trans. Geosci. Rem. Sens.*, 48 (9): 3457–3464, Sep 2010. DOI: 10.1109/TGRS.2010.2046494 Cited on page(s) 55

J. Bayliss, J. A. Gualtieri, and R. Cromp. Analysing hyperspectral data with independent component analysis. In *Proc. of SPIE*, volume 3240, pages 133–143, 1997. Cited on page(s) 92, 94

Y. Bazi and F. Melgani. Toward an optimal SVM classification system for hyperspectral remote sensing images. *IEEE Trans. Geosc. Rem. Sens.*, 44 (11): 3374–3385, Nov 2006. DOI: 10.1109/TGRS.2006.880628 Cited on page(s) 39

Y. Bazi and F. Melgani. Semisupervised PSO-SVM regression for biophysical parameter estimation. *IEEE Trans. Geosc. Rem. Sens.*, 45 (6): 1887–1895, 2007. DOI: 10.1109/TGRS.2007.895845 Cited on page(s) 107

S. G. Beaven, D. Steind, and L. E. Hoff. Comparison of Gaussian mixture and linear mixture models for classification of hyperspectral data. In *IEEE Geosc. Rem. Sens. Symp. (IGARSS)*, 2000. Cited on page(s) 73

A. Bell and T. J. Sejnowski. The 'independent components' of natural scenes are edge filters. *Vision Research*, 37: 3327–3338, 1997. DOI: 10.1016/S0042-6989(97)00121-1 Cited on page(s) 33

I.E. Bell and G.V.G. Baranoski. Reducing the dimensionality of plant spectral databases. *IEEE Trans. Geosci. Rem. Sens.*, 42 (3): 570–576, Mar 2004. DOI: 10.1109/TGRS.2003.821697 Cited on page(s) 47

R.E. Bellman. *Adaptive Control Processes: A Guided Tour*. Princeton University Press, NJ, USA, 1961. Cited on page(s) 37

J.A. Benediktsson, M. Pesaresi, and K. Arnason. Classification and feature extraction for remote sensing images from urban areas based on morphological transformations. *IEEE Trans. Geosc. Rem. Sens.*, 41: 1940–1949, 2003. DOI: 10.1109/TGRS.2003.814625 Cited on page(s) 53, 64

J.A. Benediktsson, J.A. Palmason, and J.R. Sveinsson. Classification of hyperspectral data from urban areas based on extended morphological profiles. *IEEE Trans. Geosc. Rem. Sens.*, 43: 480–490, 2005. DOI: 10.1109/TGRS.2004.842478 Cited on page(s) 53, 64

M. Berman, H. Kiiveri, R. Lagerstrom, A. Ernst, R. Dunne, and J. F Huntington. ICE: a statistical approach to identifying endmembers in hyperspectral images. *IEEE Trans. Geosc. Rem. Sens.*, 42 (10): 2085 2095, 2004. Cited on page(s) 94

J.A.J. Berni, P.J. Zarco-Tejada, L. Suárez, and E. Fereres. Thermal and narrowband multispectral remote sensing for vegetation monitoring from an unmanned aerial vehicle. *IEEE Trans. Geosc. Rem. Sens.*, 47 (3): 722– 738, 2009. DOI: 10.1109/TGRS.2008.2010457 Cited on page(s) 105

P. Bicheron and M. Leroy. A method of biophysical parameter retrieval at global scale by inversion of a vegetation reflectance model. *Rem. Sens. Environ.*, 67: 251–266, 1999. DOI: 10.1016/S0034-4257(98)00083-2 Cited on page(s) 113

J. M. Bioucas-Dias. Variable splitting augmented lagrangian approach to linear spectral unmixing. In *IEEE GRSS Workshop Hyper. Im. Sign. Proc. (WHISPERS)*, Grenoble, France, 2000. Cited on page(s) 94

J. M. Bioucas-Dias and J. Nascimento. Hyperspectral subspace identification. *IEEE Trans. Geosc. Rem. Sens.*, 46 (8): 2435–2445, 2005. DOI: 10.1109/TGRS.2008.918089 Cited on page(s) 89, 92

J. M. Bioucas-Dias and A. Plaza. Hyperspectral unmixing: Geometrical, statistical, and sparse regression-based approaches. In *SPIE Conf. Im.*

Sign. Proc. Rem. Sens., 2010. Cited on page(s) 87, 88, 93, 97

P. Birjandi and M. Datcu. Multiscale and dimensionality behavior of ICA components for satellite image indexing. *IEEE Geosc. Rem. Sens. Lett.*, 7 (1): 103–107, Jan 2010. DOI: 10.1109/LGRS.2009.2020922 Cited on page(s) 33, 54

A.R. Birks. Improvements to the AATSR IPF relating to land surface temperature retrieval and cloud clearing over land. Technical report, Rutherford Appleton Laboratory, UK, Sep 2007. Cited on page(s) 49

H. Bischof and A. Leona. Finding optimal neural networks for land use classification. *IEEE Trans. Geosci. Rem. Sens.*, 36 (1): 337–341, 1998. DOI: 10.1109/36.655348 Cited on page(s) 63

H. Bischof, W. Schneider, and A.J. Pinz. Multispectral classification of landsat-images using neural networks. *IEEE Trans. Geosci. Rem. Sens.*, 30 (3): 482–490, May 1992. DOI: 10.1109/36.142926 Cited on page(s) 64

W. J. Blackwell. *Retrieval of cloud-cleared atmospheric temperature profiles from hyperspectral infrared and microwave observations*. PhD thesis, MIT, USA, 2002. Cited on page(s) 110, 111

W.J. Blackwell. A neural-network technique for the retrieval of atmospheric temperature and moisture profiles from high spectral resolution sounding data. *IEEE Trans. Geosci. Rem. Sens.*, 43 (11): 2535–2546, Nov 2005. DOI: 10.1109/TGRS.2005.855071 Cited on page(s) 42, 111

A. Blum and P. Langley. Selection of relevant features and examples in machine learning. *Artificial Intelligence*, 97: 245–271, 1998. DOI: 10.1016/S0004-3702(97)00063-5 Cited on page(s) 38

J. Boardman. Automating spectral unmixing of AVIRIS data using convex geometry concepts. In *Summaries of the Fourth Annual JPL Airborne Geoscience Workshop, JPL Pub. 93-26 AVIRIS Workshop*, volume 1, pages 11–14, 1993. Cited on page(s) 94

J. Bobin, Y. Moudden, J.-L. Starck, and M.J. Fadili. Sparsity constraints for hyperspectral data analysis: linear mixture model and beyond. *SPIE Proceedings*, 7446, 2009. Cited on page(s) 33

J. Bolton and P. Gader. Random set framework for context-based classification with hyperspectral imagery. *IEEE Trans. Geosci. Rem. Sens.*, 47 (11): 3810–3821, Nov 2009. DOI: 10.1109/TGRS.2009.2025497 Cited on page(s) 74

J. Bolton and P. Gader. Multiple instance learning for hyperspectral image analysis. In *IEEE Geosc. Rem. Sens. Symp. (IGARSS)*, pages 4232–4235, Jul 2010. Cited on page(s) 74

J.F. Bonnans, J.C. Gilbert, C. Lemarchal,, and C.A. Sagastizábal. Numerical optimization. theoretical and practical aspects. *(Second Edition)*. *Springer, New-York. 494*, 2006. Cited on page(s) 113

C.C. Borel and S.A. Gerstl. Nonlinear spectral mixing models for vegetative and soils surface. *Remote Sensing of the Environment*, 47 (2): 403–416, 1994. DOI: 10.1016/0034-4257(94)90107-4 Cited on page(s) 86, 87

F. Bovolo. A multilevel parcel-based approach to change detection in very high resolution multi-temporal images. *IEEE Geosci. Remote Sens. Lett.*, 6 (1): 33–37, 2009. DOI: 10.1109/LGRS.2008.2007429 Cited on page(s) 68

F. Bovolo and L. Bruzzone. A split-based approach to unsupervised change detection in large size multitemporal images: application to Tsunami-damage assessment. *IEEE Trans. Geosci. Rem. Sens.*, 45 (6): 1658–1671, 2007. DOI: 10.1109/TGRS.2007.895835 Cited on page(s) 68

F. Bovolo, L. Bruzzone, and M. Marconcini. A novel approach to unsupervised change detection based on a semisupervised svm and a similarity measure. *IEEE Trans. Geosci. Rem. Sens.*, 46 (7): 2070–2082, Jul 2008. DOI: 10.1109/TGRS.2008.916643 Cited on page(s) 70

F. Bovolo, L. Bruzzone, and S. Marchesi. Analysis and adaptive estimation of the registration noise distribution in multitemporal VHR images. *IEEE*

Trans. Geosci. Rem. Sens., 47 (8): 2658–2671, 2009. DOI: 10.1109/TGRS.2009.2017014 Cited on page(s) 68

J. H. Bowles, J. A. Antoniades, M. M. Baumback, J. M. Grossmann, D. Haas, P. J. Palmadesso, and J. Stracka. Real-time analysis of hyperspectral data sets using nrl's orasis algorithm. In *Proc. of the SPIE Conference on Imaging Spectrometry III*, volume 3118, pages 38–45, 1997. Cited on page(s) 94

G.J. Briem, J.A. Benediktsson, and J.R. Sveinsson. Multiple classifiers applied to multisource remote sensing data. *IEEE Trans. Geosci. Rem. Sens.*, 40 (10): 2291–2300, 2002. DOI: 10.1109/TGRS.2002.802476 Cited on page(s) 64

J. Broadwater, A. Banerjee, and P. Burlina. Kernel methods for unmixing hyperspectral imagery. In G. Camps-Valls and L. Bruzzone, editors, *Kernel Methods for Remote Sensing Data Analysis*. Wiley & Sons, 2009. Cited on page(s) 94, 97, 99

M. Brown, H.G. Lewis, and S.R. Gunn. Linear spectral mixture models and support vector machines for remote sensing. *IEEE Trans. Geosci. Rem. Sens.*, 38 (5): 2346–2360, Sep 2000. DOI: 10.1109/36.868891 Cited on page(s) 97

L.M. Bruce, C.H. Koger, and Jiang Li. Dimensionality reduction of hyperspectral data using discrete wavelet transform feature extraction. *IEEE Trans. Geosci. Rem. Sens.*, 40 (10): 2331–2338, Oct 2002. DOI: 10.1109/TGRS.2002.804721 Cited on page(s) 42

L. Bruzzone and D. Fernández-Prieto. Unsupervised retraining of a maximum likelihood classifier for the analysis of multitemporal remote sensing images. *IEEE Trans. Geosci. Rem. Sens.*, 39 (2): 456–460, 2001. DOI: 10.1109/36.905255 Cited on page(s) 81

L. Bruzzone and F. Fernández Prieto. A technique for the selection of kernel-function parameters in RBF neural networks for classification of remote-sensing images. *IEEE Trans. Geosci. Rem. Sens.*, 37 (2): 1179–1185, 1999. DOI: 10.1109/36.752239 Cited on page(s) 64

L. Bruzzone and M. Marconcini. Toward the automatic updating of land-cover maps by a domain-adaptation SVM classifier and a circular validation strategy. *IEEE Trans. Geosci. Rem. Sens.*, 47 (4): 1108–1122, 2009. DOI: 10.1109/TGRS.2008.2007741 Cited on page(s) 82

L. Bruzzone and F. Melgani. Robust multiple estimator systems for the analysis of biophysical parameters from remotely sensed data. *IEEE Trans. Geosc. Rem. Sens.*, 43 (1): 159–174, Jan 2005. DOI: 10.1109/TGRS.2004.839818 Cited on page(s) 110

L. Bruzzone and C. Persello. A novel context-sensitive semisupervised SVM classifier robust tomislabeled training samples. *IEEE Trans. Geosci. Rem. Sens.*, 47 (7): 2142–2154, 2009. DOI: 10.1109/TGRS.2008.2011983 Cited on page(s) 82

L. Bruzzone, S. Casadio, R. Cossu, F. Sini, and C. Zehner. A system for monitoring NO2 emissions from biomass burning by using GOME and ATSR-2 data. *Int. Jour. Rem. Sens.*, 24: 1709–1721, 2003. DOI: 10.1080/01431160210144714 Cited on page(s) 111

L. Bruzzone, R. Cossu, and G. Vernazza. Detection of land-cover transitions by combining multidate classifiers. *Pattern Recogn. Lett.*, 25 (13): 1491–1500, 2004. DOI: 10.1016/j.patrec.2004.06.002 Cited on page(s) 64

L. Bruzzone, M. Chi, and M. Marconcini. A novel transductive SVM for semisupervised classification of remote sensing images. *IEEE Trans. Geosci. Rem. Sens.*, 44 (11): 3363–3373, 2006. DOI: 10.1109/TGRS.2006.877950 Cited on page(s) 76

R.W. Buccigrossi and E.P. Simoncelli. Image compression via joint statistical characterization in the wavelet domain. *IEEE Trans. on Image Proc.*, 8 (12): 1688–1701, 1999. DOI: 10.1109/83.806616 Cited on page(s) 33

P. Bunting, R. M. Lucas, K. Jones, and A. R. Bean. Characterisation and mapping of forest communities by clustering individual tree crowns.

Remote Sens. Environ., 114 (11): 2536–2547, 2010. DOI: 10.1016/j.rse.2010.05.030 Cited on page(s) 65

G. Burel. Blind separation of sources: A nonlinear neural algorithm. *Neural Networks*, 5 (6): 937–947, 1992. DOI: 10.1016/S0893-6080(05)80090-5 Cited on page(s) 43

X. Calbet and P. Schlüssel. Analytical estimation of the optimal parameters for the EOF retrievals of the IASI Level 2 product processing facility and its application using AIRS and ECMWF data. *Atmos. Chem. Phys.*, 6: 831–846, 2005. DOI: 10.5194/acp-6-831-2006 Cited on page(s) 120

J. B. Campbell. *Introduction to Remote Sensing*. The Guilford Press, New York, London, 2007. Cited on page(s) 3, 4, 11, 107

G. Camps-Valls and L. Bruzzone. Kernel-based methods for hyperspectral image classification. *IEEE Trans. Geosc. Rem. Sens.*, 43 (6): 1351–1362, Jun 2005. DOI: 10.1109/TGRS.2005.846154 Cited on page(s) 43, 64

G. Camps-Valls and L. Bruzzone. *Kernel Methods for Remote Sensing Data Analysis*. John Wiley and Sons, 2009. Cited on page(s) 15, 45, 97, 110, 116

G. Camps-Valls, L. Gómez-Chova, J. Calpe, E. Soria, J. D. Martín, L. Alonso, and J. Moreno. Robust support vector method for hyperspectral data classification and knowledge discovery. *IEEE Trans. Geosc. Rem. Sens.*, 42 (7): 1530–1542, Jul 2004. DOI: 10.1109/TGRS.2004.827262 Cited on page(s) 64

G. Camps-Valls, L. Bruzzone, J. L. Rojo-Álvarez, and F. Melgani. Robust support vector regression for biophysical variable estimation from remotely sensed images. *IEEE Geosc. Rem. Sens. Lett.*, 3 (3): 339–343, 2006a. DOI: 10.1109/LGRS.2006.871748 Cited on page(s) 110, 119

G. Camps-Valls, L. Gómez-Chova, J. Muñoz-Marí, J. Vila-Francés, and J. Calpe-Maravilla. Composite kernels for hyperspectral image classification. *IEEE Geosci. Remote Sens. Lett.*, 3 (1): 93–97, 2006b. DOI: 10.1109/LGRS.2005.857031 Cited on page(s) 64

G. Camps-Valls, L. Gómez-Chova, J. Vila-Francés, J. Amorós-López, J. Muñoz-Marí, and J. Calpe-Maravilla. Retrieval of oceanic chlorophyll concentration with relevance vector machines. *Rem. Sens. Environ.*, 105 (1): 23–33, Nov 2006c. DOI: 10.1016/j.rse.2006.06.004 Cited on page(s) 110, 119

G. Camps-Valls, T.V. Bandos Marsheva, and D. Zhou. Semi-supervised graph-based hyperspectral image classification. *IEEE Trans. Geosci. Rem. Sens.*, 45 (10): 3044–3054, 2007. DOI: 10.1109/TGRS.2007.895416 Cited on page(s) 76

G. Camps-Valls, L. Gómez-Chova, J. Muñoz-Marí, J. Luis Rojo-Álvarez, and M. Martínez-Ramón. Kernel-based framework for multi-temporal and multi-source remote sensing data classification and change detection. *IEEE Trans. Geosc. Rem. Sens.*, 46 (6): 1822–1835, 2008a. DOI: 10.1109/TGRS.2008.916201 Cited on page(s) 70, 71, 74

G. Camps-Valls, J. Gutiérrez, G. Gómez, and J. Malo. On the suitable domain for SVM training in image coding. *Journal of Machine Learning Research*, 9 (1): 49–66, 2008b. Cited on page(s) 29, 34

G. Camps-Valls, J. Muñoz Marí, L. Gómez-Chova, K. Richter, and J. Calpe-Maravilla. Biophysical parameter estimation with a semi-supervised support vector machine. *IEEE Geosc. Rem. Sens. Lett.*, 6 (2): 248–252, Feb 2009. DOI: 10.1109/LGRS.2008.2009077 Cited on page(s) 107

G. Camps-Valls, J. Mooij, and B. Scholkopf. Remote sensing feature selection by kernel dependence measures. *IEEE Geosc. Rem. Sens. Lett.*, 7 (3): 587–591, Jul 2010a. Cited on page(s) 39, 40

G. Camps-Valls, N. Shervashidze, and K.M. Borgwardt. Spatio-spectral remote sensing image classification with graph kernels. *IEEE Geosci.Rem.Sens.Lett.*, 7 (4): 741–745, 2010b. DOI: 10.1109/LGRS.2010.2046618 Cited on page(s) 20, 55

G. Camps-Valls, J. Muñoz Marí, L. Gómez-Chova, L. Guanter, and X. Calbet. Nonlinear statistical retrieval of atmospheric profiles from

MetOp-IASI and MTG-IRS infrared sounding data. *IEEE Trans. Geosc. Rem. Sens.*, 49, 2011. DOI: 10.1109/TGRS.2011.2168963 Cited on page(s) 111, 116, 120

M. J. Canty and A. A. Nielsen. Automatic radiometric normalization of multitemporal satellite imagery with the iteratively re-weighted MAD transformation. *Remote Sens. Environ.*, 112 (3): 1025–1036, 2008. DOI: 10.1016/j.rse.2007.07.013 Cited on page(s) 69

L. Capobianco, A. Garzelli, and G. Camps-Valls. Target detection with semisupervised kernel orthogonal subspace projection. *IEEE Trans. Geosci. Rem. Sens.*, 47 (11): 3822–3833, Jul 2009. DOI: 10.1109/TGRS.2009.2020910 Cited on page(s) 74, 76

P. Capolsini, S. Andréfouët, C. Rion, and C. Payri. A comparison of Landsat ETM+, SPOT HRV, Ikonos, ASTER, and airborne MASTER data for coral reef habitat mapping in South Pacific islands. *Can. J. Remote Sensing*, 29 (2): 187–200, 2003. DOI: 10.5589/m02-088 Cited on page(s) 12

M.J. Carlotto. A cluster-based approach for detecting man-made objects and changes in imagery. *IEEE Trans. Geosci. Rem. Sens.*, 43 (2): 374–387, Feb 2005. DOI: 10.1109/TGRS.2004.841481 Cited on page(s) 73

T. Celik. Multiscale change detection in multitemporal satellite images. *IEEE Geosci. Remote Sens. Lett.*, 6 (4): 820–824, Oct 2009. DOI: 10.1109/LGRS.2009.2026188 Cited on page(s) 69

G. Chalon, F. Cayla, and D. Diebel. IASI: an advanced sounder for operational meteorology. In *Proceedings of the 52nd Congress of IAF*, Toulouse, France, 1-5 October 2001, 2001. Cited on page(s) 21, 22

P.K. Chan and B.C. Gao. A comparison of MODIS, NCEP, and TMI sea surface temperature datasets. *IEEE Geoscience Remote Sensing Letters*, 2 (3): 270–274, 2005. DOI: 10.1109/LGRS.2005.846838 Cited on page(s) 109

T. Chan, C. Chi, Y. Huang, and W. Ma. Convex analysis based minimum-volume enclosing simplex algorithm for hyperspectral unmixing,. *IEEE*

Transactions on Signal Processing, 57 (11): 4418–4432, 2009. DOI: 10.1109/TSP.2009.2025802 Cited on page(s) 94

C.-I. Chang. *Hyperspectral Imaging: Techniques for Spectral Detection and Classification*. Plenum Publishing Co., 2003. Cited on page(s) 72, 83

C.-I Chang and Q. Du. Estimation of number of spectrally distinct signal sources in hyperspectral imagery. *IEEE Trans. Geosc. Rem. Sens.*, 42 (3): 608–619, 2004. DOI: 10.1109/TGRS.2003.819189 Cited on page(s) 89, 92

C. I. Chang and D. C. Heinz. Constrained subpixel target detection for remotely sensed imagery. *IEEE Trans. Geosc. Rem. Sens.*, 38 (3): 1144–1159, 2000. DOI: 10.1109/36.843007 Cited on page(s) 72

C.-I. Chang and B. Ji. Weighted abundance-constrained linear spectral mixture analysis. *IEEE Trans. Geosc. Rem. Sens.*, 44 (2): 378–388, Feb 2006. DOI: 10.1109/TGRS.2005.861408 Cited on page(s) 97

C.-I. Chang, C.-C. Wu, W. Liu, and Y.-C. Ouyang. A new growing method for simplex-based endmember extraction algorithm. *IEEE Trans. Geosc. Rem. Sens.*, 44 (11): 2804–2819, 2006. DOI: 10.1109/TGRS.2006.881803 Cited on page(s) 94

S. Chang, M. J. Westfield, F. Lehmann, D. Oertel, and R. Richter. A 79–channel airborne imaging spectrometer. *Proc. SPIE*, 1937: 164–172, 1993. DOI: 10.1117/12.157053 Cited on page(s) 21, 23, 25

C. Chen and X. Zhang. Independent component analysis for remote sensing study. In *SPIE Conf. Im. Sign. Proc. Rem. Sens.*, volume 3871, pages 150–158, 1999. Cited on page(s) 92

G. Chen and S.-E. Qian. Denoising of hyperspectral imagery using principal component analysis and wavelet shrinkage. *IEEE Trans. Geosci. Rem. Sens.*, 49 (3): 973–980, 2011. DOI: 10.1109/TGRS.2010.2075937 Cited on page(s) 34, 54

J. Chen, X. Chen, X. Cui, and J. Chen. Change vector analysis in posterior probability space: A new method for land cover change detection. *IEEE*

Geosci. Remote Sens. Lett., 8 (2): 317–321, 2010a. DOI: 10.1109/LGRS.2010.2068537 Cited on page(s) 68, 69

S. Chen and D. Zhang. Semisupervised dimensionality reduction with pairwise constraints for hyperspectral image classification. *IEEE Geosc. Rem. Sens. Lett.*, 8 (2): 369–373, Mar 2011. DOI: 10.1109/LGRS.2010.2076407 Cited on page(s) 42

S. Chen, D. Donoho, and M. Saunders. Atomic decomposition by basis pursuit. *SIAM Review*, 43 (1): 129–159, 2001. DOI: 10.1137/S003614450037906X Cited on page(s) 94

X. Chen, T. Fang, H. Huo, and D. Li. Semisupervised feature selection for unbalanced sample sets of VHR images. *IEEE Geosc. Rem. Sens. Lett.*, 7 (4): 781–785, Oct 2010b. DOI: 10.1109/LGRS.2010.2048197 Cited on page(s) 39

M. Chi and L. Bruzzone. Semisupervised classification of hyperspectral images by SVMs optimized in the primal. *IEEE Trans. Geosc. Rem. Sens.*, 45 (6): 1870–1880, 2007. DOI: 10.1109/TGRS.2007.894550 Cited on page(s) 76

P. Cipollini, G. Corsini, M. Diani, and R. Grass. Retrieval of sea water optically active parameters from hyperspectral data by means of generalized radial basis function neural networks. *IEEE Trans. Geosc. Rem. Sens.*, 39: 1508–1524, 2001. DOI: 10.1109/36.934081 Cited on page(s) 110

R.J. Clarke. Relation between the Karhunen Loève and cosine transforms. *IEE Proc*, 128 (6): 359–360, 1981. Cited on page(s) 20, 21, 25

D.A. Clausi. Comparison and fusion of co-occurrence, Gabor and MRF texture features for classification of SAR sea-ice imagery. *Atmosphere-Ocean*, 39 (3): 183–194, Mar 2001. DOI: 10.1080/07055900.2001.9649675 Cited on page(s) 51

D.A. Clausi and H. Deng. Design-based texture feature fusion using Gabor filters and co-occurrence probabilities. *IEEE Transactions on Image*

Processing, 14 (7): 925–936, Jul 2005. DOI: 10.1109/TIP.2005.849319 Cited on page(s) 50, 51

D.A. Clausi and B. Yue. Comparing cooccurrence probabilities and Markov random fields for texture analysis of SAR sea ice imagery. *IEEE Trans. Geosci. Rem. Sens.*, 42 (1): 215–228, Jan 2004. DOI: 10.1109/TGRS.2003.817218 Cited on page(s) 50, 51

B. Combal. Retrieval of canopy biophysical variables from bi-directional reflectance data. Using prior information to solve the ill-posed inverse problem. *Rem. Sens. Environ.*, 84: 1–15, 2002. DOI: 10.1016/S0034-4257(02)00035-4 Cited on page(s) 113, 114

B. Combal, F. Baret, and M. Weiss. Improving canopy variables estimation from remote sensing data by exploiting ancillary information. case study on sugar beet canopies. *Agronomie 22*, pages 2–15, 2001. Cited on page(s) 103

P. Comon. Independent component analysis: A new concept? *Signal Processing*, 36 (3): 287–314, 1994. DOI: 10.1016/0165-1684(94)90029-9 Cited on page(s) 33, 42

R.G. Congalton and K. Green. *Assessing the Accuracy of Remotely Sensed Data: Principles and Practices*. CRC Press, Boca Raton, FL, USA, 1 edition, 1999. Cited on page(s) 40

N. C. Coops, M-L. Smith, M.E. Martin, and S. V. Ollinger. Prediction of eucalypt foliage nitrogen content from satellite-derived hyperspectral data. *IEEE Trans. Geosc. Rem. Sens.*, 41 (6): 1338–1346, Jun 2003. DOI: 10.1109/TGRS.2003.813135 Cited on page(s) 107

P. Coppin, I. Jonckheere, K. Nackaerts, and Muys B. Digital change detection methods in ecosystem monitoring: A review. *Int. J. Remote Sens*, 25 (9): 1565–1596, 2004. DOI: 10.1080/0143116031000101675 Cited on page(s) 67, 69

J. Crespo, J. Serra, and R. Schafer. Theoretical aspects of morphological filters by reconstruction. *Signal Processing*, 47: 201–225, 1995. DOI: 10.1016/0165-1684(95)00108-5 Cited on page(s) 52

G.R. Cross and A.K. Jain. Markov random field texture models. *IEEE Trans. on Pattern Analysis and Machine Intelligence*, PAMI-5 (1): 25–39, 1983. DOI: 10.1109/TPAMI.1983.4767341 Cited on page(s) 51

M. A. Cutter. Compact high-resolution imaging spectrometer (CHRIS) design and performance. In Sylvia S. Shen and Paul E. Lewis, editors, *Imaging Spectrometry X*, volume 5546, pages 126–131. SPIE, 2004. Cited on page(s) 14

M. Dalla Mura, J. A. Benediktsson, F. Bovolo, and L. Bruzzone. An unsupervised technique based on morphological filters for change detection in very high resolution images. *IEEE Trans. Geosci. Rem. Sens.*, 5 (3): 433–437, 2008. DOI: 10.1109/LGRS.2008.917726 Cited on page(s) 68

M. Dalla Mura, J.A. Benediktsson, B. Waske, and L. Bruzzone. Morphological attribute profiles for the analysis of very high resolution images. *IEEE Trans. Geosci. Rem. Sens.*, 48 (10): 3747–3762, Oct 2010. Cited on page(s) 53

G. A. D'Almeida, P. Koepke, and E. P. Shettle. *Atmospheric Aerosol, Global Climatology and Radiative characteristics*. A. Deepak, Hampton, USA, 1991. Cited on page(s) 7, 8

F. M. Danson and S. E. Plummer. *Advances in Environmental Remote Sensing*. John Wiley & Sons, New York, 1995. Cited on page(s) 2

R. Darvishzadeh, A. Skidmore, M. Schlerf, and C. Atzberger. Inversion of a radiative transfer model for estimating vegetation LAI and chlorophyll in a heterogeneous grassland. *Rem. Sens. Environ.*, 112 (5): 2592–2604, 2008. DOI: 10.1016/j.rse.2007.12.003 Cited on page(s) 114

P. Dash, F.-M. Gottsche, F.-S. Olesen, and H. Fischer. Land surface temperature and emissivity estimation from passive sensor data: theory and pratice–current trends. *Int. J. Rem. Sens.*, 23 (13): 2563–2594, 2002. DOI: 10.1080/01431160110115041 Cited on page(s) 109

P. Debba, E.J.M. Carranza, F.D. van der Meer, and A. Stein. Abundance estimation of spectrally similar minerals by using derivative spectra in simulated annealing. *IEEE Trans. Geosc. Rem. Sens.*, 44 (12): 3649–3658, Dec 2006. DOI: 10.1109/TGRS.2006.881125 Cited on page(s) 97

F. Del Frate and G. Schiavon. Nonlinear principal component analysis for the radiometric inversion of atmospheric profiles by using neural networks. *IEEE Trans. Geosci. Rem. Sens.*, 37 (5): 2335–2342, Sep 1999. DOI: 10.1109/36.789630 Cited on page(s) 43

F. Dell'Acqua, P. Gamba, A. Ferrari, J.A. Palmason, J.A. Benediktsson, and K. Arnason. Exploiting spectral and spatial information in hyperspectral urban data with high resolution. *IEEE Geosci. Rem. Sens. Lett.*, 1 (4): 322–326, 2004. DOI: 10.1109/LGRS.2004.837009 Cited on page(s) 25

B. Demir, C. Persello, and L. Bruzzone. Batch mode active learning methods for the interactive classification of remote sensing images. *IEEE Trans. Geosci. Rem. Sens.*, 49 (3): 1014–1032, 2011. DOI: 10.1109/TGRS.2010.2072929 Cited on page(s) 78, 79, 80

W. Di and M. M. Crawford. Active learning via multi-view and local proximity co-regularization for hyperspectral image classification. *IEEE J. Sel. Topics Signal Proc.*, 5 (3): 618–628, 2011. DOI: 10.1109/JSTSP.2011.2123077 Cited on page(s) 78, 80

R. Dianat and S. Kasaei. Dimension reduction of optical remote sensing images via minimum change rate deviation method. *IEEE Trans. Geosci. Rem. Sens.*, 48 (1): 198–206, Jan 2010. DOI: 10.1109/TGRS.2009.2024306 Cited on page(s) 43, 55

E. Doi, T. Inui, T. Lee, T. Wachtler, and T. Sejnowski. Spatiochromatic receptive field properties derived from information-theoretic analysis of cone mosaic responses to natural scenes. *Neural Computation*, 15 (2): 397–417, 2003. DOI: 10.1162/089976603762552960 Cited on page(s) 25, 26

L. David Donoho. De-noising by soft-thresholding. *IEEE Trans. Information Theory*, 41 (3): 613–627, 1995. DOI: 10.1109/18.382009 Cited on page(s) 34

I. Dopido, M. Zortea, A. Villa, A. Plaza, and P. Gamba. Unmixing prior to supervised classification of remotely sensed hyperspectral images. *IEEE Geosc. Rem. Sens. Lett.*, 8 (4): 760–764, Jul 2011. Cited on page(s) 85

P.L. Dragotti, G. Poggi, and A.R.P. Ragozini. Compression of multispectral images by three-dimensional SPIHT algorithm. *IEEE Trans. Geosci. Rem. Sens.*, 38 (1-II): 416–428, 2000. DOI: 10.1109/36.823937 Cited on page(s) 34

M. Dundar and A. Langrebe. A cost-effective semisupervised classifier approach with kernels. *IEEE Trans. Geosci. Rem. Sens.*, 42 (1): 264–270, Jan 2004. DOI: 10.1109/TGRS.2003.817815 Cited on page(s) 76

S.S. Durbha, R.L. King, and N.H. Younan. Support vector machines regression for retrieval of leaf area index from multiangle imaging spectroradiometer. *Rem. Sens. Environ.*, 107 (1-2): 348–361, 2007. DOI: 10.1016/j.rse.2006.09.031 Cited on page(s) 107, 116

B. Dzwonkowski and X.-H. Yan. Development and application of a neural network based colour algorithm in coastal waters. *Int. Jour. Rem. Sens.*, 26 (6): 1175–1200, Mar 2005. DOI: 10.1080/01431160512331326549 Cited on page(s) 110

R.W. Ehrich. Detection of global edges in textured images. *Computers, IEEE Transactions on*, C-26 (6): 589–603, Jun 1977. Cited on page(s) 50

M. Elad, M.A.T. Figueiredo, and Yi Ma. On the role of sparse and redundant representations in image processing. *Proceedings of the IEEE*, 98 (6): 972–982, Jun 2010. DOI: 10.1109/JPROC.2009.2037655 Cited on page(s) 21, 33

A.J Elmore, J.F Mustard, S.J Manning, and D.B. Lobell. Quantifying vegetation change in semiarid environments: Precision and accuracy of spectral mixture analysis and the normalized difference vegetation index. *Rem. Sens. Environ.*, 73 (1): 87–102, 2000. DOI: 10.1016/S0034-4257(00)00100-0 Cited on page(s) 85

I. Epifanio and P. Soille. Morphological texture features for unsupervised and supervised segmentations of natural landscapes. *IEEE Trans. Geosc.*

Rem. Sens., 45: 1074–1083, 2007. DOI: 10.1109/TGRS.2006.890581 Cited on page(s) 53

I. Epifanio, J. Gutiérrez, and J. Malo. Linear transform for simultaneous diagonalization of covariance and perceptual metric matrix in image coding. *Pattern Recognition*, 36 (8): 1799–1811, 2003. DOI: 10.1016/S0031-3203(02)00325-4 Cited on page(s) 29

European Space Agency. MERIS Product Handbook. Technical Report 2.1, European Space Agency, Oct 2006. Cited on page(s) 49

European Space Agency. AATSR Product Handbook. Technical Report 2.2, European Space Agency, Feb 2007. Cited on page(s) 49

H. Fang and S Liang. Retrieving LAI from Landsat 7 ETM+ data with a neural network method: Simulation and validation study. *IEEE Trans. Geosc. Rem. Sens.*, 41: 2052–2062, 2003. Cited on page(s) 112, 113

H. Fang and S. Liang. A hybrid inversion method for mapping leaf area index from MODIS data: Experiments and application to broadleaf and needleleaf canopies. *Rem. Sens. Environ.*, 94 (3): 405–424, 2005. DOI: 10.1016/j.rse.2004.11.001 Cited on page(s) 115

H. Fang, S. Liang, and A. Kuusk. Retrieving leaf area index using a genetic algorithm with a canopy radiative transfer model. *Rem. Sens. Environ.*, 85: 257–270, 2003. DOI: 10.1016/S0034-4257(03)00005-1 Cited on page(s) 115

T.A. Farmer, C.J.Q. andNelson, M.A. Wulder, and C. Derksen. Identification of snow cover regimes through spatial and temporal clustering of satellite microwave brightness temperatures. *Remote Sens. Environ.*, 114 (1): 199–210, 2010. DOI: 10.1016/j.rse.2009.09.002 Cited on page(s) 65

H. Fattahi, M.J.V. Zoej, M.R. Mobasheri, M. Dehghani, and M.R. Sahebi. Windowed fourier transform for noise reduction of SAR interferograms. *IEEE Geosc. Rem. Sens. Lett.*, 6 (3): 418–422, Jul 2009. DOI: 10.1109/LGRS.2009.2015338 Cited on page(s) 54

M. Fauvel, J. Chanussot, and J.A. Benediktsson. Decision fusion for the classification of urban remote sensing images. *IEEE Trans. Geosci. Rem. Sens.*, 44 (10): 2828–2838, 2006. DOI: 10.1109/TGRS.2006.876708 Cited on page(s) 64

M. Fauvel, J.A. Benediktsson, J. Chanussot, and J.R. Sveinsson. Spectral and spatial classification of hyperspectral data using SVMs and morphological profiles. *IEEE Trans. Geosc. Rem. Sens.*, 46 (11): 3804–3814, 2008. DOI: 10.1109/TGRS.2008.922034 Cited on page(s) 50, 53, 64

M. Ferecatu and N. Boujemaa. Interactive remote sensing image retrieval using active relevance feedback. *IEEE Trans. Geosci. Rem. Sens.*, 45 (4): 818–826, 2007. DOI: 10.1109/TGRS.2007.892007 Cited on page(s) 79, 80

M. D. Fleming, J. S. Berkebile, and R. M. Hoffer. Computer-aided analysis of LANDSAT-I MSS data: a comparison of three approaches, including a "modified clustering" approach. LARS information note 072475, Purdue University, 1975. Cited on page(s) 81

J. Font, G.S.E. Lagerloef, D.M. Le Vine, A. Camps, and O.-Z. Zanife. The determination of surface salinity with the european SMOS space mission. *IEEE Trans. Geosc. Rem. Sens.*, 42 (10): 2196–2205, Oct 2004. DOI: 10.1109/TGRS.2004.834649 Cited on page(s) 108

G. M. Foody. Thematic map comparison: Evaluating the statistical significance of differences in classification accuracy. *Photogramm. Eng. Rem. S.*, 50 (5): 627–633, 2004. Cited on page(s) 61, 63

G. M. Foody and J. Mathur. A relative evaluation of multiclass image classification by support vector machines. *IEEE Trans. Geosci. Rem. Sens.*, pages 1–9, Jul 2004. Cited on page(s) 64

P. S. Frazier and K. L. Page. Water body detection and delineation with Landsat TM data. *Photogrammetric engineering and remote sensing*, 66: 1461–1465, 2000. Cited on page(s) 108

M. A. Friedl and C. E. Brodley. Decision tree classification of land cover from remotely sensed data. *Remote Sens. Environ.*, 61: 399–409, 1997. DOI: 10.1016/S0034-4257(97)00049-7 Cited on page(s) 63

H. Frigui and P. Gader. Detection and discrimination of land mines in ground-penetrating radar based on edge histogram descriptors and a possibilistic-nearest neighbor classifier. *IEEE Trans. Geosci. Rem. Sens.*, 17 (1): 185–199, Feb 2009. Cited on page(s) 74

S. Fukuda and H. Hirosawa. A wavelet-based texture feature set applied to classification of multifrequency polarimetric SAR images. *IEEE Trans. Geosci. Rem. Sens.*, 37 (5): 2282–2286, Sep 1999. DOI: 10.1109/36.789624 Cited on page(s) 54

K. Fukunaga. *Introduction to Statistical Pattern Classification*. Academic Press, San Diego, California, USA, 1990. Cited on page(s) 42

K. Fukunaga and R.R. Hayes. Effects of sample size in classifier design. *IEEE Trans. on Pattern Analysis and Machine Intelligence*, 11 (8): 873–885, 1989. DOI: 10.1109/34.31448 Cited on page(s) 38

R. Furfaro, R. D. Morris, A. Kottas, M. Taddy, and B. D. Ganapol. A Gaussian Process Approach to Quantifying the Uncertainty of Vegetation Parameters from Remote Sensing Observations. *AGU Fall Meeting Abstracts*, pages A261+, Dec 2006. Cited on page(s) 107

P.D. Gader, M. Mystkowski, and Yunxin Zhao. Landmine detection with ground penetrating radar using hidden markov models. *IEEE Trans. Geosci. Rem. Sens.*, 39 (6): 1231–1244, June 2001. DOI: 10.1109/36.927446 Cited on page(s) 74

F. J. García-Haro, M. A. Gilabert, and J. Melia. Linear spectral mixture modeling to estimate vegetation amount from optical spectral data. *Int. J. Rem. Sens.*, 17: 3373–3400, 1996. DOI: 10.1080/014311169608949157 Cited on page(s) 86

F.J. García-Haro, S. Sommer, and T. Kemper. Variable multiple endmember spectral mixture analysis (VMESMA). *Int. Jour. Rem. Sens.*, 26 (10):

2135–216, 2005. DOI: 10.1080/01431160512331337817 Cited on page(s) 93

F. García-Vílchez, J. Muñoz-Marí, M. Zortea, I. Blanes-García, V. González, G. Camps-Valls, A. Plaza, and J. Serra-Sagristà. On the impact of lossy compression on hyperspectral image classification and unmixing. *IEEE Geosc. Rem. Sens. Lett.*, 8 (2): 253–257, 2011. DOI: 10.1109/LGRS.2010.2062484 Cited on page(s) 21, 93

S. A. Garver and D. A. Siegel. Inherent optical property inversion of ocean color spectra and its biogeochemical interpretation. Time series from the Sargasso Sea. *Journal of Geophysical Research*, 102: 18607–18625, 1997. DOI: 10.1029/96JC03243 Cited on page(s) 109

F. Gemmell, J. Varjo, M. Strandstrom, and A. Kuusk. Comparison of measured boreal forest characteristics with estimates from tm data and limited ancillary information using reflectance model inversion. *Rem. Sens. Environ.*, 81 (2-3): 365–377, 2002. DOI: 10.1016/S0034-4257(02)00012-3 Cited on page(s) 113

A. Gersho and R. M. Gray. *Vector Quantization and Signal Compression*. Kluwer Academic Press, Boston, 1992. Cited on page(s) 22, 27, 28, 29, 32

S. Ghosh, L. Bruzzone, S. Patra, F. Bovolo, and A. Ghosh. A context-sensitive technique for unsupervised change detection based on Hopfield-type neural networks. *IEEE Trans. Geosci. Rem. Sens.*, 45 (3): 778–789, 2007. DOI: 10.1109/TGRS.2006.888861 Cited on page(s) 69

D. Gillis, J. Bowles, G. M. Lamela, W. J. Rhea, C. M. Bachmann, M. Montes, and T. Ainsworth. Manifold learning techniques for the analysis of hyperspectral ocean data. In S. S. Shen and P. E. Lewis, editors, *Proc. of the SPIE conferece on Algorithms and Technologies for Multispectral, Hyperspectral and Ultraspectral Imagery XI*, volume 5806, pages 342–351, 2005. Cited on page(s) 89

N.S. Goel. *Inversion of canopy reflectance models for estimation of biophysical parameters from reflectance data*, pages 205–251. Wiley,

New York, 1989. Cited on page(s) 112

A.F.H. Goetz, G. Vane, J. Solomon, and B.N. Rock. Imaging spectrometry for Earth remote sensing. *Science*, 228: 1147–1153, 1985. DOI: 10.1126/science.228.4704.1147 Cited on page(s) 12, 19

D. E Goldberg. Genetic algorithms in search, optimization and machine learning. *Addison-Wesley*, 1989. Cited on page(s) 113

E.H. Gombrich. *The Story of Art*. Phaidon Press, London, 16th edition, 1995. Cited on page(s) 19

L. Gómez-Chova, G. Camps-Valls, J. Calpe, L. Guanter, and J. Moreno. Cloud-screening algorithm for ENVISAT/MERIS multispectral images. *IEEE Trans. Geosc. Rem. Sens.*, 45 (12, Part 2): 4105–4118, Dec 2007. DOI: 10.1109/TGRS.2007.905312 Cited on page(s) 49

L. Gómez-Chova, L. Alonso, L. Guanter, G. Camps-Valls, J. Calpe, and J. Moreno. Correction of systematic spatial noise in push-broom hyperspectral sensors: application to CHRIS/PROBA images. *Appl. Opt.*, 47 (28): F46–F60, Oct 2008. DOI: 10.1364/AO.47.000F46 Cited on page(s) 21, 116

L. Gómez-Chova, G. Camps-Valls, J. Muñoz-Marí, and J. Calpe. Semi-supervised image classification with laplacian support vector machines. *IEEE Geosci. Remote Sens. Lett.*, 5 (4): 336–340, 2008. DOI: 10.1109/LGRS.2008.916070 Cited on page(s) 76

L. Gómez-Chova, G. Camps-Valls, L. Bruzzone, and J. Calpe-Maravilla. Mean map kernel methods for semisupervised cloud classification. *IEEE Trans. Geosci. Rem. Sens.*, 48 (1): 207–220, 2010. DOI: 10.1109/TGRS.2009.2026425 Cited on page(s) 70, 82

L. Gómez-Chova, R. Jenssen, and G. Camps-Valls. Kernel Entropy Component Analysis for Remote Sensing Image Clustering. *IEEE Geoscience and Remote Sensing Letters*, 2011a. Cited on page(s) 43

L. Gómez-Chova, J. Muñoz-Marí, V. Laparra, J. Malo-López, and G. Camps-Valls. *Optical Remote Sensing. Advances in Signal Processing and Exploitation Techniques*, chapter A Review of Kernel Methods in Remote Sensing Data Analysis, pages 171–206. Springer, Germany, Feb 2011b. Cited on page(s) 43

L. Gómez-Chova, A.A. Nielsen, and G. Camps-Valls. Explicit signal to noise ratio in reproducing kernel Hilbert spaces. In *IEEE Geosc. Rem. Sens. Symp. (IGARSS)*, pages 3570–3570. IEEE, Jul 2011c. Cited on page(s) 43, 46

P. Gong, S. X. Wang, and S Liang. Inverting a canopy reflectance model using a neural network. *Int. J. Rem. Sens.*, 20: 111–122, 1999. DOI: 10.1080/014311699213631 Cited on page(s) 115

R.C. González and R.E. Woods. *Digital Image Processing (3rd Edition)*. Prentice Hall, Aug 2007. ISBN 013168728X. Cited on page(s) 49, 50

M. González-Audicana, J.L. Saleta, R.G. Catalan, and R. Garcia. Fusion of multispectral and panchromatic images using improved IHS and PCA mergers based on wavelet decomposition. *IEEE Trans. Geosci. Rem. Sens.*, 42 (6): 1291–1299, Jun 2004. DOI: 10.1109/TGRS.2004.825593 Cited on page(s) 54

M.C. González-Sanpedro, T. Le Toan, J. Moreno, L. Kergoat, and E. Rubio. Seasonal variations of leaf area index of agricultural fields retrieved from Landsat data. *Rem. Sens. Environ.*, 112 (3): 810–824, 2008. DOI: 10.1016/j.rse.2007.06.018 Cited on page(s) 114

N. Goodwin, N.C. Coops, and C. Stone. Assessing plantation canopy condition from airborne imagery using spectral mixture analysis and fractional abundances. *International Journal of Applied Earth Observation and Geoinformation*, 7 (1): 11–28, 2005. DOI: 10.1016/j.jag.2004.10.003 Cited on page(s) 85

S. Gopal and C. Woodcock. Remote sensing of forest change using artificial neural networks. *IEEE Trans. Geosci. Remote Sens*, 34: 398–404, 1996. DOI: 10.1109/36.485117 Cited on page(s) 115

M. Graña, I. Villaverde, J. O. Maldonado, and C. Hernández. Two lattice computing approaches for the unsupervised segmentation of hyperspectral images. *Neurocomputing*, 72: 2111–2120, 2009. DOI: 10.1016/j.neucom.2008.06.026 Cited on page(s) 94

A. Green, M. Berman, P. Switzer, and M. D. Craig. A transformation for ordering multispectral data in terms of image quality with implications for noise removal. *IEEE Trans. Geosc. Rem. Sens.*, 26 (1): 65–74, 1988a. DOI: 10.1109/36.3001 Cited on page(s) 89

A.A. Green, M. Berman, P. Switzer, and M.D. Craig. A transformation for ordering multispectral data in terms of image quality with implications for noise removal. *IEEE Trans. Geosci. Rem. Sens.*, 26 (1): 65–74, Jan 1988b. DOI: 10.1109/36.3001 Cited on page(s) 42

R.O. Green, G. Asner, S. Ungar, and R. Knox. Results of the Decadal Survey HyspIRI Imaging Spectrometer Concept Study: A High Signal-To-Noise Ratio and High Uniformity Global Mission to Measure Plant Physiology and Functional Type. In *IEEE Geosc. Rem. Sens. Symp. (IGARSS)*, Boston, USA, Jul 2008. Cited on page(s) 14

J. Gruninger, A. Ratkowski, and M. Hoke. The sequential maximum angle convex cone (SMACC) endmember model. In *Proceedings of SPIE*, volume 5425, 2004. Cited on page(s) 94

Y. Gu, Y. Liu, and Y. Zhang. A selective KPCA algorithm based on high-order statistics for anomaly detection in hyperspectral imagery. *IEEE Geosci. Remote Sens. Letters*, 5 (1): 43–47, Jan 2008. Cited on page(s) 43, 73

L. Guanter, L. Alonso, and J. Moreno. A method for the surface reflectance retrieval from proba/chris data over land: Application to ESA SPARC campaigns. *IEEE Trans. Geosc. Rem. Sens.*, 43 (12): 2908–2917, 2005. DOI: 10.1109/TGRS.2005.857915 Cited on page(s) 116

L. Guanter, L. Gómez-Chova, and J. Moreno. Coupled retrieval of aerosol optical thickness, columnar water vapor and surface reflectance maps from ENVISAT/MERIS data over land. *Rem. Sens. Environ.*, 112 (6): 2898–2913, Jun 2008. DOI: 10.1016/j.rse.2008.02.001 Cited on page(s) 15, 49

J. Gutiérrez, F. J. Ferri, and J. Malo. Regularization operators for natural images based on nonlinear perception models. *IEEE Trans. Image Proc.*, 15 (1): 189–200, January 2006. DOI: 10.1109/TIP.2005.860345 Cited on page(s) 33, 34

Isabelle Guyon, Jason Weston, Stephen Barnhill, and Vladimir Vapnik. Gene selection for cancer classification using support vector machines. *Machine Learning*, 46 (1-3), 2002. DOI: 10.1023/A:1012487302797 Cited on page(s) 40

D. Haboudane, J. R. Miller, E. Pattey, P. J. Zarco-Tejada, and I. B. Strachan. Hyperspectral vegetation indices and novel algorithms for predicting green LAI of crop canopies: Modeling and validation in the context of precision agriculture. *Rem. Sens. Environ.*, 90 (3): 337–352, 2004. DOI: 10.1016/j.rse.2003.12.013 Cited on page(s) 105

D. Haboudane, N. Tremblay, J.R. Miller, and P. Vigneault. Remote estimation of crop chlorophyll content using spectral indices derived from hyperspectral data. *IEEE Trans. Geosc. Rem. Sens.*, 46 (2): 423–436, 2008. DOI: 10.1109/TGRS.2007.904836 Cited on page(s) 105

S.L. Haines, G.J. Jedlovec, and S.M. Lazarus. A MODIS sea surface temperature composite for regional applications. *IEEE Trans. Geosc. Rem. Sens.*, 45 (9): 2919–2927, 2007. DOI: 10.1109/TGRS.2007.898274 Cited on page(s) 109

J. Ham, D. D. Lee, S. Mika, and B. Schölkopf. A kernel view of the dimensionality reduction of manifolds. In *Intern. Conf. Mach. Learn.*, ICML'04, pages 47–55, 2004. Cited on page(s) 43

P.J.B. Hancock, R.J. Baddeley, and L.S. Smith. The principal components of natural images. *Network*, 3: 61–72, 1992. DOI: 10.1088/0954-898X/3/1/008 Cited on page(s) 23

M. Hansen, R. Dubayah, and R. Defries. Classification trees: an alternative to traditional land cover classifiers. *Int. J. Rem. Sens.*, 17 (5): 1075–1081, 1996. DOI: 10.1080/01431169608949069 Cited on page(s) 63

B. Hapke. Bidirection reflectance spectroscopy. I. Theory. *J. of Geophysical Research*, 86: 3039–3054, 1981. DOI: 10.1029/JB086iB04p03039 Cited on page(s) 86, 87

B. Hapke. *Theory of Reflectance and Emittance Spectroscopy*. Cambridge University Press, 1993. Cited on page(s) 7

R.M. Haralick, K. Shanmugam, and I.H. Dinstein. Textural features for image classification. *Systems, Man and Cybernetics, IEEE Transactions on*, 3 (6): 610–621, Nov 1973. Cited on page(s) 51

N. J. Hardman-Mountford, T. Hirata, K. A. Richardson, and J. Aiken. An objective methodology for the classification of ecological pattern into biomes and provinces for the pelagic ocean. *Remote Sens. Environ.*, 112 (8): 3341–3352, 2008. DOI: 10.1016/j.rse.2008.02.016 Cited on page(s) 65

J. Harsanyi, W. Farrand, and C.-I. Chang. Determining the number and identity of spectral endmembers: An integrated approach using Neyman-Pearson eigenthresholding and iterative constrained RMS error minimization. In *Proc. 9th Thematic Conf. Geologic Remote Sensing*, 1993. Cited on page(s) 89, 92

J.C. Harsanyi and C.I. Chang. Hyperspectral image classification and dimensionality reduction: an orthogonal subspace projection approach. *IEEE Trans. Geosci. Rem. Sens.*, 32 (4): 779–785, Jul 1994. DOI: 10.1109/36.298007 Cited on page(s) 42, 97

M. Hasanzadeh and S. Kasaei. A multispectral image segmentation method using size-weighted fuzzy clustering and membership connectedness. *IEEE Geosci. Remote Sens. Lett.*, 7 (3): 520–524, 2010. DOI: 10.1109/LGRS.2010.2040800 Cited on page(s) 65

M. Hassner and J. Sklansky. The use of Markov Random Fields as models of texture. *Computer Graphics and Image Processing*, 12 (4): 357–370, 1980. DOI: 10.1016/0146-664X(80)90019-2 Cited on page(s) 51

E. Hecht. *Optics*. Addison Wesley, Boston, MA, 4th edition, 2001. Cited on page(s) 19, 21

D. Heinz and C.-I Chang. Fully constrained least squares linear mixture analysis for material quantification in hyperspectral imagery. *IEEE Trans. Geosc. Rem. Sens.*, 39: 529–545, 2000. DOI: 10.1109/36.911111 Cited on page(s) 97

T. Hilker, N. C. Coops, M. A. Wulder, T. A. Black, and R. D. Guy. The use of remote sensing in light use efficiency based models of gross primary production: A review of current status and future requirements. *Science of the Total Environment*, 404 (2-3): 411–423, 2008. DOI: 10.1016/j.scitotenv.2007.11.007 Cited on page(s) 104

F. Hilton, N. C. Atkinson, S. J. English, and J. R. Eyre. Assimilation of IASI at the Met Office and assessment of its impact through observing system experiments. *Q. J. R. Meteorol. Soc.*, 135: 495–505, 2009. DOI: 10.1002/qj.379 Cited on page(s) 120

R. Houborg, H. Soegaard, and E. Boegh. Combining vegetation index and model inversion methods for the extraction of key vegetation biophysical parameters using terra and aqua modis reflectance data. *Rem. Sens. Environ.*, 106 (1): 39–58, 2007. DOI: 10.1016/j.rse.2006.07.016 Cited on page(s) 113

C. Huang, L. S. Davis, and J. R. G. Townshend. An assessment of support vector machines for land cover classification. *Int. J. Rem. Sens.*, 23 (4): 725–749, 2002. DOI: 10.1080/01431160110040323 Cited on page(s) 64

H. L. Huang, W. L. Smith, and H. M. Woolf. Vertical resolution and accuracy of atmospheric infrared sounding spectrometers. *J. Appl. Meteor.*, 31: 265–274, 1992. DOI: 10.1175/1520-0450(1992)031%3C0265:VRAAOA%3E2.0.CO;2 Cited on page(s) 120

G.F. Hughes. On the mean accuracy of statistical pattern recognizers. *IEEE Trans. Inf. Theory*, 14 (1): 55–63, 1968. DOI: 10.1109/TIT.1968.1054102 Cited on page(s) 37

T. Hultberg. EUMETSAT technical report: IASI principal component compression.

http://www.eumetsat.int/Home/Main/DataProducts/Resources/index .htm, 2009. Cited on page(s) 29

A. Hyvärinen. Sparse code shrinkage: Denoising of non-gaussian data by ML estimation. *Neural Computation*, 11: 1739–1768, 1999. DOI: 10.1162/089976699300016214 Cited on page(s) 22, 27, 34

A. Hyvärinen and P. Pajunen. Nonlinear independent component analysis: Existence and uniqueness results. *Neural Networks*, 12 (3): 429–439, 1999. DOI: 10.1016/S0893-6080(98)00140-3 Cited on page(s) 43

A. Hyvärinen, J. Karhunen, and E. Oja. *Independent Component Analysis*. John Wiley & Sons, New York, 2001. Cited on page(s) 33

A. Hyvarinen, J. Hurri, and J. Vayrynen. Bubbles: A unifying framework for low-level statistical properties of natural image sequences. *JOSA A*, 20 (7): 1237–1252, 2003. DOI: 10.1364/JOSAA.20.001237 Cited on page(s) 33

A. Ifarraguerri and C.-I Chang. Multispectral and hyperspectral image analysis with convex cones. *IEEE Trans. Geosc. Rem. Sens.*, 37 (2): 756–770, 1999. DOI: 10.1109/36.752192 Cited on page(s) 94

A. Ifarraguerri and C.-I. Chang. Unsupervised hyperspectral image analysis with projection pursuit. *IEEE Trans. Geosc. Rem. Sens.*, 38 (6): 127–143, 2000. Cited on page(s) 89

J. Im, J. R. Jensen, and M. E. Hodgson. Optimizing the binary discriminant function in change detection applications. *Remote Sens. Environ.*, 112 (6): 2761–2776, 2008. DOI: 10.1016/j.rse.2008.01.007 Cited on page(s) 68

J. Inglada, V. Muron, D. Pichard, and T. Feuvrier. Analysis of artifacts in subpixel remote sensing image registration. *IEEE Trans. Geosc. Rem. Sens.*, 45 (1): 254–264, 2007. DOI: 10.1109/TGRS.2006.882262 Cited on page(s) 21

M-D. Iordache, J. M. Bioucas-Dias, and A. Plaza. Sparse Unmixing of Hyperspectral Data. *IEEE Trans. Geosc. Rem. Sens.*, 49 (6): 2014–2039,

2011. DOI: 10.1109/TGRS.2010.2098413 Cited on page(s) 93

Q. Jackson and D. Landgrebe. An adaptive classifier design for high-dimensional data analysis with a limited training data set. *IEEE Trans. Geosci. Rem. Sens.*, 39 (12): 2664–2679, 2001. DOI: 10.1109/36.975001 Cited on page(s) 76

S. Jacquemoud and F. Baret. PROSPECT: a model of leaf optical properties spectra. *Rem. Sens. Environ.*, 34: 75–91, 1990. DOI: 10.1016/0034-4257(90)90100-Z Cited on page(s) 112

S. Jacquemoud, C. Bacour, H. Poilvé, and J.-P. Frangi. Comparison of four radiative transfer models to simulate plant canopies reflectance: Direct and inverse mode. *Rem. Sens. Environ.*, 74 (3): 471–481, 2000. DOI: 10.1016/S0034-4257(00)00139-5 Cited on page(s) 112, 113

S. Jacquemoud, W. Verhoef, F. Baret, C. Bacour, P.J. Zarco-Tejada, G.P. Asner, C. François, and S.L. Ustin. PROSPECT + SAIL models: A review of use for vegetation characterization. *Rem. Sens. Environ.*, 113 (1): S56–S66, 2009. DOI: 10.1016/j.rse.2008.01.026 Cited on page(s) 111, 112

Y. Jin and C. Liu. Biomass retrieval from high dimensional active/passive remote data by using artificial neural networks. *Int. J. Remote Sensing*, 18: 971–979, 1997. DOI: 10.1080/014311697218863 Cited on page(s) 115

I.T. Jollife. *Principal Component Analysis*. Springer-Verlag, 1986. Cited on page(s) 42, 89

A. Journel and C. Huijbregts. *Mining Geostatistics*. Academic Press, 1978. Cited on page(s) 54

G. Jun and J. Ghosh. An efficient active learning algorithm with knowledge transfer for hyperspectral remote sensing data. In *IEEE Geosc. Rem. Sens. Symp. (IGARSS)*, volume 1, pages I–52–55, Boston, USA, 2008. Cited on page(s) 79

D. L. B. Jupp, K. K. Mayo, D. A. Kucker, D. Van, R. Classen, R. A. Kenchinton, and P. R. Guerin. Remote sensing for planning and managing the great barrier reef of Australia. *Photogrammetria*, 40: 21–42, 1985. DOI: 10.1016/0031-8663(85)90043-2 Cited on page(s) 109

U. Kandaswamy, D.A. Adjeroh, and M.C. Lee. Efficient texture analysis of SAR imagery. *IEEE Trans. Geosci. Rem. Sens.*, 43 (9): 2075–2083, Sep 2005. DOI: 10.1109/TGRS.2005.852768 Cited on page(s) 50

L. D. Kapla. Inference of atmospheric structure from remote radiantion measurements. *JOSA*, 49: 100–1, 1959. Cited on page(s) 110

H. Kaufmann, L. Guanter, K. Segl, S. Hofer, K.-P. Foerster, T. Stuffler, A. Müller, R. Richter, H. Bach, P. Hostert, and C. Chlebek. Environmental Mapping and Analysis Program (EnMAP) – Recent Advances and Status. In *IEEE Geosc. Rem. Sens. Symp. (IGARSS)*, Boston, USA, Jul 2008. DOI: 10.1109/IGARSS.2008.4779668 Cited on page(s) 14

L. E. Keiner. Estimating oceanic chlorophyll concentrations with neural networks. *Int. Jour. Rem. Sens.*, 20 (1): 189–194, Jan 1999. DOI: 10.1080/014311699213695 Cited on page(s) 110

Y.H. Kerr, P. Waldteufel, J.-P. Wigneron, J. Martinuzzi, J. Font, and M. Berger. Soil moisture retrieval from space: the soil moisture and ocean salinity (smos) mission. *IEEE Trans. Geosc. Rem. Sens.*, 39 (8): 1729–1735, Aug 2001. DOI: 10.1109/36.942551 Cited on page(s) 108

N. Keshava and J. F. Mustard. Spectral unmixing. *IEEE Sign. Proc. Mag.*, 19 (1): 44–57, 2002. DOI: 10.1109/79.974727 Cited on page(s) 83, 84, 86, 88

N. Keshava, J. Kerekes, D. Manolakis, and G. Shaw. An algorithm taxonomy for hyperspectral unmixing. In *Proc. of the SPIE AeroSense Conference on Algorithms for Multispectral and Hyperspectral Imagery VI*, volume 4049, pages 42–63, 2000. Cited on page(s) 92

S. Khorram. Water quality mapping from Lansat digital data. *Int. Jour. Rem. Sens.*, 2: 143–153, 1980. Cited on page(s) 109

D.S. Kimes, R.F. Nelson, M.T. Manry, and A.K. Fung. Attributes of neural networks for extracting continuous vegetation variables from optical and radar measurements. *Int. Jour. Rem. Sens.*, 19 (14): 2639–2663, 1998. DOI: 10.1080/014311698214433 Cited on page(s) 115

D.S. Kimes, Y. Knyazikhin, J.L. Privette, A.A. Abuelgasim, and F. Gao. Inversion methods for physically-based models. *Remote Sens. Rev.*, 18: 381–439, 2000. DOI: 10.1080/02757250009532396 Cited on page(s) 112

D.S. Kimes, J. Gastellu-Etchegorry, and P. Estève. Recovery of forest canopy characteristics through inversion of a complex 3D model. *Rem. Sens. Environ.*, 79 (2-3): 320–328, 2002. DOI: 10.1016/S0034-4257(01)00282-6 Cited on page(s) 115

J. I. L. King. *The radiative heat transfer of planet Earth*, page 133. Univ. Michigan Press, 1956. Cited on page(s) 110

Y. Knyazikhin. Estimation of vegetation canopy leaf area index and fraction of absorbed photosynthetically active radiation from atmosphere-corrected MISR data. *J. Geophys. Res.*, 103 (D24): 32239–32256, 1998a. DOI: 10.1029/98JD02461 Cited on page(s) 114

Y. Knyazikhin. MODIS leaf area index (LAI), and fraction of photosynthetically active radiation absorbed by vegetation FPAR. Technical report, GSFC/NASA, http://eospso.gsfc.nasa.gov/atbd/modistables.html, 1999. Cited on page(s) 103, 114

Y. Knyazikhin, J.V. Martonchik, R.B. Myneni, D.J. Diner, and S.W. Running. Synergistic algorithm for estimating vegetation canopy leaf area index and fraction of absorbed photosynthetically active radiation from MODIS and MISR data. *J. Geophys. Res.103(D24)*, pages 32257–32275, 1998. DOI: 10.1029/98JD02462 Cited on page(s) 114

R. Kohavi and G.H. John. Wrappers for features subset selection. *Int. J. Digit. Libr.*, 1: 108–121, 1997. Cited on page(s) 38

A. Kokhanovsky,W. von Hoyningen-Huene, J.P. Burrows, O. Colin, J.M. Rosaz, and E. Mathot. The Determination of the Cloud Fraction in the

SCIAMACHY Pixel using MERIS. In H. Lacoste and L. Ouwehand, editors, *Proceedings of the 2nd MERIS/(A)ATSR Workshop*, pages CD–Rom, Frascati, Italy, Nov 2008. ESA SP-666, ESA Publications Division. Cited on page(s) 49

M.A. Kramer. Nonlinear principal component analysis using autoassociative neural networks. *AIChE Journal*, 37 (2): 233–243, 1991. DOI: 10.1002/aic.690370209 Cited on page(s) 43

S. Kraut, L.L. Scharf, and L.T. McWhorter. Adaptive subspace detectors. *Signal Processing, IEEE Transactions on*, 49 (1): 1–16, Jan 2001. Cited on page(s) 74

P. Kruizinga, N. Petkov, and S.E. Grigorescu. Comparison of texture features based on Gabor filters. *Image Analysis and Processing, 1999. Proceedings. International Conference on*, pages 142–147, 1999. Cited on page(s) 50

M. P. Krystek. An algorithm to calculate correlated colour temperature. *Color Res. Appl.*, 10 (1): 38–40, 1985. DOI: 10.1002/col.5080100109 Cited on page(s) 22

B.C. Kuo and D.A. Landgrebe. Nonparametric weighted feature extraction for classification. *IEEE Trans. Geosci. Rem. Sens.*, 42 (5): 1096–1105, May 2004. DOI: 10.1109/TGRS.2004.825578 Cited on page(s) 42

B.C. Kuo, C.H. Li, and J.M. Yang. Kernel nonparametric weighted feature extraction for hyperspectral image classification. *IEEE Trans. Geosci. Rem. Sens.*, 47 (4): 1139–1155, Apr 2009. DOI: 10.1109/TGRS.2008.2008308 Cited on page(s) 43

E.J. Kwiatkowska and G.S. Fargion. Application of machine-learning techniques toward the creation of a consistent and calibrated global chlorophyll concentration baseline dataset using remotely sensed ocean color data. *IEEE Trans. Geosc. Rem. Sens.*, 41 (12): 2844–2860, Dec 2003. DOI: 10.1109/TGRS.2003.818016 Cited on page(s) 110

H. Kwon and N. Nasrabadi. Kernel RX-algorithm: a nonlinear anomyla detector for hyperspectral imagery. *IEEE Trans. Geosci. Rem. Sens.*, 43

(2): 388–397, 2005. DOI: 10.1109/TGRS.2004.841487 Cited on page(s) 73, 74

H. Kwon and N. Nasrabadi. A comparative analysis of kernel subspace target detectors for hyperspectral imagery. *EURASIP Journal of of Advances in Signal Proc.*, 2007 (29250), 2007a. DOI: 10.1155/2007/29250 Cited on page(s) 72, 74

H. Kwon and N. Nasrabadi. Kernel spectral matched filter for hyperspectral imagery. *Intl. J. Computer Vision*, 71 (2): 127–141, 2007b. DOI: 10.1007/s11263-006-6689-3 Cited on page(s) 74

H. Kwon and N. M. Nasrabadi. Kernel adaptive subspace detector for hyperspectral imagery. *IEEE Geosc. Rem. Sens. Lett.*, 3 (2): 271–275, Apr 2006a. DOI: 10.1109/LGRS.2006.869985 Cited on page(s) 74

H. Kwon and N. M. Nasrabadi. Kernel matched subspace detectors for hyperspectral target detection. *IEEE Trans. Patt. Anal. Mach. Intell.*, 28 (2): 178–194,Feb 2006b. DOI: 10.1109/TPAMI.2006.39 Cited on page(s) 74

G.S.E. Lagerloef, C. T. Swift, and D. M. Le Vine. Sea surface salinity: the next remote sensing challenge. *Oceanography*, 8 (2): 44–50, 1995. Cited on page(s) 108

V. Laparra, J. Muñoz-Marí, G. Camps-Valls, and J. Malo. PCA Gaussianization for one-class remote sensing image classification. In *Proceedings of SPIE–The International Society for Optical Engineering*, volume 7477, 2009. Cited on page(s) 43

V. Laparra, J. Gutiérrez, G. Camps-Valls, and J. Malo. Image denoising with kernels based on natural image relations. *Journal of Machine Learn. Res.*, 11: 873–903, Feb 2010. Cited on page(s) 33, 34

V. Laparra, G. Camps, and J. Malo. Iterative Gaussianization: from ICA to random rotations. *IEEE Trans. Neur. Nets.*, 22 (4): 537–549, 2011a. DOI: 10.1109/TNN.2011.2106511 Cited on page(s) 20

V. Laparra, D. Tuia, S. Jiménez, G. Camps-Valls, and Jesús Malo. Principal polynomial analysis for remote sensing data processing. In *IEEE International Conference Geoscience and Remote Sensing*, 2011b. Cited on page(s) 32

G. Le Maire, C. Francois, K. Soudani, D. Berveiller, J. Y. Pontailler, N. Breda, H. Genet, H. Davi, and E. Dufrane. Calibration and validation of hyperspectral indices for the estimation of broadleaved forest leaf chlorophyll content, leaf mass per area, leaf area index and leaf canopy biomass. *Rem. Sens. Environ.*, 112 (10): 3846–3864, 2008. DOI: 10.1016/j.rse.2008.06.005 Cited on page(s) 105

C. Lee and D.A. Landgrebe. Feature extraction based on decision boundaries. *IEEE Trans. Patt. Anal. Mach. Intell.*, 15 (4): 388–400, Apr 1993. DOI: 10.1109/34.206958 Cited on page(s) 42

J. B. Lee, S. Woodyatt, and M. Berman. Enhancement of high spectral resolution remote-sensing data by noise-adjusted principal components transform. *IEEE Trans. Geosc. Rem. Sens.*, 28 (3): 295–304, 1990. DOI: 10.1109/36.54356 Cited on page(s) 89

S. Lee. Efficient multistage approach for unsupervised image classification. In *IEEE Geosc. Rem. Sens. Symp. (IGARSS)*, pages 1581–1584, Anchorage, AK, USA, 2004. Cited on page(s) 65

S. Lee and M. M Crawford. Hierarchical clustering approach for unsupervised image classification of hyperspectral data. In *IEEE Geosc. Rem. Sens. Symp. (IGARSS)*, pages 941–944, Anchorage, AK, USA, 2004. Cited on page(s) 65

Z. P. Lee, K. L. Carder, and R. Arnone. Deriving inherent optical properties from water color: A multi-band quasi-analytical algorithm for optically deep waters. *Applied Optics*, 41: 5755–5772, 2002. DOI: 10.1364/AO.41.005755 Cited on page(s) 109

J. M. Leiva-Murillo, L. Gómez-Chova, and G. Camps-Valls. Multitask SVM learning for remote sensing data classification. In *Proceedings of the SPIE Remote Sensing Conference*, Toulouse, France, 2010. Cited on page(s) 82

M. Lennon, M. Mouchot, G. Mercier, and L. Hubert-Moy. Independent component analysis as a tool for the dimensionality reduction and the representation of hyperspectral images. In *Proc. of the IEEE Int. Geoscience and Remote Sensing Symp*, 2001. Cited on page(s) 89

D. Letexier and S. Bourennane. Noise removal from hyperspectral images by multidimensional filtering. *IEEE Trans. Geosci. Rem. Sens.*, 46 (7): 2061–2069, 2008. DOI: 10.1109/TGRS.2008.916641 Cited on page(s) 34

B. Li, R. Yang, and H. Jiang. Remote-sensing image compression using two-dimensional oriented wavelet transform. *IEEE Trans. Geosci. Rem. Sens.*, 49 (1): 236–250, Jan 2011. DOI: 10.1109/TGRS.2010.2056691 Cited on page(s) 50, 54

J. Li and R.M. Narayanan. Integrated spectral and spatial information mining in remote sensing imagery. *IEEE Trans. Geosci. Rem. Sens.*, 42 (3): 673–685, Mar 2004. DOI: 10.1109/TGRS.2004.824221 Cited on page(s) 54, 97

J. Li, J.M. Bioucas-Dias, and A. Plaza. Semisupervised hyperspectral image segmentation using multinomial logistic regression with active learning. *IEEE Trans. Geosci. Rem. Sens.*, 48 (11): 4085–4098, 2010. Cited on page(s) 79

L. Li, S.L. Ustin, and D. Riaño. Retrieval of fresh leaf fuel moisture content using genetic algorithm partial least squares (GA-PLS) modeling. *IEEE Geosc. Rem. Sens. Lett.*, 4 (2): 216–220, Apr 2007. DOI: 10.1109/LGRS.2006.888847 Cited on page(s) 107

Q. Li and Z. Wang. Reduced-reference image quality assessment using divisive normalization-based image representation. *IEEE J. Sel. Topics Sig. Proc.*, 3 (2): 202–211, 2009. Cited on page(s) 21

S. Liang. *Quantitative Remote Sensing of Land Surfaces*. John Wiley & Sons, New York, 2004. Cited on page(s) 2, 3, 4, 48, 101, 103, 105, 106, 112

S. Liang. *Advances in Land Remote Sensing: System, Modeling, Inversion and Applications*. Springer Verlag, Germany, 2008. Cited on page(s) 4, 101, 103, 106, 112, 113

S. Liang, H. Fang, M. Kaul, T. G. Van Niel, T. R. Mcvicdr, J. Pearlman, C. L. Walthall, C. Daughtry, and K. F. Huemmrich. Estimation of land surface broadband albedos and leaf area index from EO-1 ALI data and validation. *IEEE Trans. Geosci. Rem. Sens.*, 41: 1260–1268, 2003. Cited on page(s) 115

G. Licciardi, F. Del Frate, and R. Duca. Feature reduction of hyperspectral data using autoassociative neural networks algorithms. In *IEEE Geosc. Rem. Sens. Symp. (IGARSS)*, volume 1, pages 176–179, Jul 2009a. Cited on page(s) 43

G. Licciardi, F. Pacifici, D. Tuia, S. Prasad, T. West, F. Giacco, J. Inglada, E. Christophe, J. Chanussot, and P. Gamba. Decision fusion for the classification of hyperspectral data: Outcome of the 2008 GRS-S data fusion contest. *IEEE Trans. Geosci. Rem. Sens.*, 47 (11): 3857–3865, 2009b. DOI: 10.1109/TGRS.2009.2029340 Cited on page(s) 65

H. K. Lichtenthaler. Chlorophylls and carotenoids: Pigments of photosynthetic biomembranes. *Methods Enzymol.*, 148: 350–382, 1987. DOI: 10.1016/0076-6879(87)48036-1 Cited on page(s) 104

T. M. Lillesand, R. W. Kiefer, and J. Chipman. *Remote Sensing and Image Interpretation*. John Wiley & Sons, New York, 2008. Cited on page(s) 2, 3, 4, 11, 101

Y.-C. Lin and K. Sarabandi. Retrieval of forest parameters using a fractal-based coherent scattering model and a genetic algorithm. *IEEE Trans. Geosc. Rem. Sens.*, 37 (3): 1415–1424, May 1999. Cited on page(s) 113

R. Lindstrot, R. Preusker, and J. Fischer. The retrieval of land surface pressure from MERIS measurements in the oxygen A band. *Journal of Atmospheric and Oceanic Technology*, 26: 1367–1377, 2009. DOI: 10.1175/2009JTECHA1212.1 Cited on page(s) 49

K. N. Liou. *An Introduction to Atmospheric Radiation*. Academic Press, Hampton, USA, second edition, 2002. Cited on page(s) 120

A. Liu, G. Jun, and J. Ghosh. Active learning with spatially sensitive labeling costs. In *NIPS Workshop on Cost-sensitive Learning*, 2008a. Cited on page(s) 79

J. Liu and P. Moulin. Information-theoretic analysis of interscale and intrascale dependencies between image wavelet coefficients. *IEEE Trans. Image Proc.*, 10 (11): 1647–1658, 2001. DOI: 10.1109/83.967393 Cited on page(s) 33

Q. Liu, X. Liao, and L. Carin. Detection of unexploded ordnance via efficient semisupervised and active learning. *IEEE Trans. Geosci. Rem. Sens.*, 46 (9): 2558–2567, 2008b. Cited on page(s) 79

D. Llewellyn-Jones, M. C. Edwards, C. T. Mutlow, A. R. Birks, I. J. Barton, and H. Tait. AATSR: global-change and surface-temperature measurements from Envisat. *ESA Bulletin*, 105: 11–21, February 2001. Cited on page(s) 41

S.P. Lloyd. Least squares quantization in PCM. *IEEE Trans. Inform. Theory*, 28 (2): 129–137, 1982. DOI: 10.1109/TIT.1982.1056489 Cited on page(s) 29

D.B. Lobell and G.P. Asner. Cropland distributions from temporal unmixing of MODIS data. *Rem. Sens. Environ.*, 93 (3): 412–422, 2004. DOI: 10.1016/j.rse.2004.08.002 Cited on page(s) 85

A. Lorette, X. Descombes, and J. Zerubia. Texture analysis through a Markovian modelling and fuzzy classification: Application to urban area extraction from satellite images. *Int. J. Comput. Vision*, 36 (3): 221–236, 2000. DOI: 10.1023/A:1008129103384 Cited on page(s) 50, 51

A. Lotsch, M.A. Friedl, and J. Pinzon. Spatio-temporal deconvolution of NDVI image sequences using independent component analysis. *IEEE Trans. Geosci. Rem. Sens.*, 41 (12): 2938–2942, Dec 2003. DOI: 10.1109/TGRS.2003.819868 Cited on page(s) 42

G. Louverdis, M. I. Vardavoulia, I. Andreadis, and P. Tsalides. New approach to morphological color image processing. *Pattern Recognition*, 35 (8): 1733–1741, 2002. DOI: 10.1016/S0031-3203(01)00166-2 Cited on page(s) 53

X. Lu, L.E. Hoff, I.S. Reed, M. Chen, and L.B. Stotts. Automatic target detection detection and recognition in multiband imagery: a unified ML detection and estimation approach. *IEEE Trans. Image Proc.*, 6: 143–156, 1997. DOI: 10.1109/83.552103 Cited on page(s) 72, 74

T. Luo, K. Kramer, D. B. Goldgof, L. O. Hall, S. Samson, A. Remsen, and T. Hopkins. Active learning to recognize multiple types of plankton. *J. Mach. Learn. Res.*, 6: 589–613, 2005. Cited on page(s) 78, 79, 80

L. Ma, M.M. Crawford, and J. Tian. Local manifold learning-based-nearest-neighbor for hyperspectral image classification. *IEEE Trans. Geosci. Rem. Sens.*, 48 (11): 4099–4109, Nov 2010. Cited on page(s) 43

D. Makowski, J. Hillier, D. Wallach, B. Andrieu, and M.H. Jeuffroy. *Parameter estimation for crop models*, pages 101–149. Elsevier, Amsterdam, 2006. Cited on page(s) 114

W. A. Malila. Change vector analysis: An approach for detecting forest change with landsat. In *IEEE Proceedings of Annual Symposium on Machine Processing of Remotely Sensing Data*, pages 326–336, 1980. Cited on page(s) 68

J. Malo and V. Laparra. Psychophysically tuned divisive normalization approximately factorizes the PDF of natural images. *Neural Computation*, 22 (12): 3179–3206, 2011. DOI: 10.1162/NECO_a_00046 Cited on page(s) 33

J. Malo, F. Ferri, J. Albert, J. Soret, and J. Artigas. The role of perceptual contrast non-linearities in image transform coding. *Image & Vision Comp.*, 18 (3): 233–246, 2000. DOI: 10.1016/S0262-8856(99)00010-4 Cited on page(s) 22, 27, 28, 29

J. Malo, I. Epifanio, R. Navarro, and E. Simoncelli. Non-linear image representation for efficient perceptual coding. *IEEE Trans. Image Proc.*,

15 (1): 68–80, 2006. DOI: 10.1109/TIP.2005.860325 Cited on page(s) 29, 33, 34

L.T. Maloney. Evaluation of linear models of surface spectral reflectance with small numbers of parameters. *JOSA A*, 3 (10): 1673–1683, 1986. DOI: 10.1364/JOSAA.3.001673 Cited on page(s) 23

A. Marcal and L. Castro. Hierarchical clustering of multispectral images using combined spectral and spatial criteria. *IEEE Geosci. Remote Sens. Lett.*, 2 (1): 59–63, 2005. DOI: 10.1109/LGRS.2004.839646 Cited on page(s) 65

S. Maritorena and J.E. O'Reilly. *OC2v2: Update on the initial operational SeaWiFS chlorophyll algorithm*, volume 11, pages 3–8. John Wiley & Sons, NASA Goddard Space Flight Center, Greenbelt, Maryland, USA, 2000. Cited on page(s) 109, 118

P. M. Mather. *Computer Processing of Remotely-Sensed Images: An Introduction*. John Wiley & Sons, New York, 2004. Cited on page(s) 4

U. Maulik and S. Bandyopadhyay. Fuzzy partitioning using a real-coded variable-length genetic algorithm for pixel classification. *IEEE Trans. Geosci. Rem. Sens.*, 41 (5): 1075–1081, 2003. DOI: 10.1109/TGRS.2003.810924 Cited on page(s) 64

U. Maulik and I. Saha. Modified differential evolution based fuzzy clustering for pixel classification in remote sensing imagery. *Pattern Recogn.*, 42 (9): 2135–2149, 2009. DOI: 10.1016/j.patcog.2009.01.011 Cited on page(s) 64

S.K. Meher, B.U. Shankar, and A. Ghosh. Wavelet-feature-based classifiers for multispectral remote-sensing images. *IEEE Trans. Geosci. Rem. Sens.*, 45 (6): 1881–1886, June 2007. DOI: 10.1109/TGRS.2007.895836 Cited on page(s) 54

F. Melgani and L. Bruzzone. Classification of hyperspectral remote sensing images with support vector machines. *IEEE Trans. Geosci. Rem. Sens.*, 42 (8): 1778–1790, 2004. DOI: 10.1109/TGRS.2004.831865 Cited on page(s) 64

L. Miao and H. Qi. Endmember extraction from highly mixed data using minimum volume constrained nonnegative matrix factorization. *IEEE Trans. Geosc. Rem. Sens.*, 45 (3): 765–777, 2007. DOI: 10.1109/TGRS.2006.888466 Cited on page(s) 94

L. Miao, H. Qi, and H. Szu. A maximum entropy approach to unsupervised mixed-pixel decomposition. *IEEE Trans. Im. Proc.*, 16 (4): 1008–1021, 2007. DOI: 10.1109/TIP.2006.891350 Cited on page(s) 97

P. Mitra, B. Uma Shankar, and S.K. Pal. Segmentation of multispectral remote sensing images using active support vector machines. *Pattern Recogn. Lett.*, 25 (9): 1067–1074, 2004. DOI: 10.1016/j.patrec.2004.03.004 Cited on page(s) 78, 79, 80

A. Mohan, G. Sapiro, and E. Bosch. Spatially coherent nonlinear dimensionality reduction and segmentation of hyperspectral images. *IEEE Geosc. Rem. Sens. Lett.*, 4 (2): 206–210, Apr 2007. DOI: 10.1109/LGRS.2006.888105 Cited on page(s) 43

B. Mojaradi, H. Abrishami-Moghaddam, M.J.V. Zoej, and R.P.W. Duin. Dimensionality reduction of hyperspectral data via spectral feature extraction. *IEEE Trans. Geosci. Rem. Sens.*, 47 (7): 2091–2105, Jul 2009. DOI: 10.1109/TGRS.2008.2010346 Cited on page(s) 42

J. Moncet, G. Uymin, A. E. Lipton, and H. E. Snell. Infrared radiance modeling by optimal spectral sampling. *J. Atmos. Sci.*, 65: 3917–3934, 2008. DOI: 10.1175/2008JAS2711.1 Cited on page(s) 112, 120

J.A. Morgan. Bayesian estimation for land surface temperature retrieval: the nuisance of emissivities. *IEEE Trans. Geosc. Rem. Sens.*, 43 (6): 1279–1288, Jun 2005. DOI: 10.1109/TGRS.2005.845637 Cited on page(s) 114

H. Mott. *Remote Sensing with Polarimetric Radar*. John Wiley & Sons, New York, 2007. Cited on page(s) 2

G. Mountrakis, J. Im, and C. Ogole. Support vector machines in remote sensing: A review. *ISPRS Journal of Photogrammetry and Remote Sensing*, 66 (3): 247–259, 2011. DOI: 10.1016/j.isprsjprs.2010.11.001 Cited on page(s) 107

J.P. Muller and L. Brinckmann. Stereo analysis tools for cloud-top heights and cloud masks from (A)ATSR(2). In H. Lacoste and L. Ouwehand, editors, *Proceedings of the 2nd MERIS/(A)ATSR Workshop*, pages CD–Rom, Frascati, Italy, Nov 2008. ESA SP-666, ESA Publications Division. Cited on page(s) 49

J. Muñoz-Marí, L. Bruzzone, and G. Camps-Vails. A support vector domain description approach to supervised classification of remote sensing images. *IEEE Trans. Geosci. Rem. Sens.*, 45 (8): 2683–2692, Aug 2007. DOI: 10.1109/TGRS.2007.897425 Cited on page(s) 74

J. Muñoz-Marí, A. Plaza, J.A. Gualtieri, and G. Camps-Valls. Parallel programming and applications in grid, P2P and networking systems. In F. Xhafa, editor, *Parallel Implementation of SVM in Earth Observation Applications*. IOS Press, UK, 2009. Cited on page(s) 64

J. Muñoz-Marí, F. Bovolo, L. Gómez-Chova, L. Bruzzone, and G. Camps-Valls. Semisupervised one-class support vector machines for classification of remote sensing data. *IEEE Trans. Geosci. Rem. Sens.*, 48 (8): 3188–3197, 2010. DOI: 10.1109/TGRS.2010.2045764 Cited on page(s) 70, 74, 76

R. B. Myneni, F. G. Hall, P. J. Sellers, and A. L. Marshak. Interpretation of spectral vegetation indexes. *IEEE Trans. Geosc. Rem. Sens.*, 33 (2): 481–486, 1995. DOI: 10.1109/36.377948 Cited on page(s) 105, 112

R.B. Myneni, S. Hoffman, Y. Knyazikhin, J.L Privette, J. Glassy, Y. Tian, Y. Wang, X Song, Y. Zhang, G.R. Smith, A. Lotsch, M. Friedl, J.T. Morisette, P. Votava, R.R. Nemani, and S.W. Running. Global products of vegetation leaf area and fraction absorbed PAR from year one of MODIS data. *Rem. Sens. Environ.*, 83 (1-2): 214–231, 2002. DOI: 10.1016/S0034-4257(02)00074-3 Cited on page(s) 114

E. Naesset, O. Martin Bollandsas, and T. Gobakken. Comparing regression methods in estimation of biophysical properties of forest stands from two different inventories using laser scanner data. *Rem. Sens. Environ.*, 94 (4): 541–553, Feb 2005. DOI: 10.1016/j.rse.2004.11.010 Cited on page(s) 107

J. Nascimento and J. M. Bioucas-Dias. Vertex component analysis: a fast algorithm to unmix hyperspectral data. *IEEE Trans. Geosc. Rem. Sens.*, 43 (8): 898–910, 2005a. Cited on page(s) 91, 94

J. Nascimento and J. M. Bioucas-Dias. Hyperspectral unmixing algorithm via dependent component analysis. In *IEEE Geosc. Rem. Sens. Symp. (IGARSS)*, Barcelona, Spain, Jul 2007. Cited on page(s) 94

J. Nascimento and J.M. Bioucas-Dias. Does independent component analysis play a role in unmixing hyperspectral data? *IEEE Trans. Geosci. Rem. Sens.*, 43 (1): 175–187, Jan 2005b. DOI: 10.1109/TGRS.2004.839806 Cited on page(s) 42, 92, 95

N. Nasrabadi. Target detection with kernels. In G. Camps-Valls and L. Bruzzone, editors, *Kernel methods for remote sensing data analysis*, pages 147–168. J. Wiley & Sons, NJ, USA, 2009. Cited on page(s) 74

R.A Neville, K Staenz, T Szeredi, J Lefevbre, and P Hauff. Automatic endmember extraction from hyperspectral data for mineral exploration. In *International Airborne Remote Sensing Conference and Exhibition, 4th/21st Canadian Symposium on Remote Sensing*, Ottawa, Canada, 1999. Cited on page(s) 94

A. A. Nielsen. The regularized iteratively reweighted MAD method for change detection in multi-and hyperspectral data. *IEEE Trans. Image Process.*, 16 (2): 463–478, 2006. DOI: 10.1109/TIP.2006.888195 Cited on page(s) 68

A. A. Nielsen, K. Conradsen, and J. J. Simpson. Multivariate alteration detection (MAD) and MAF post-processing in multispectral, bitemporal image data: New approaches to change detection studies. *Remote Sens. Environ.*, 64: 1–19, 1998. DOI: 10.1016/S0034-4257(97)00162-4 Cited on page(s) 68

A.A. Nielsen. Kernel maximum autocorrelation factor and minimum noise fraction transformations. *IEEE Trans. Image Processing*, 20 (3): 612–624, Mar 2011. DOI: 10.1109/TIP.2010.2076296 Cited on page(s) 43

C. Notarnicola, M. Angiulli, and F. Posa. Soil moisture retrieval from remotely sensed data: Neural network approach versus bayesian method. *IEEE Trans. Geosc. Rem. Sens.*, 46 (2): 547–557, Jan 2008. DOI: 10.1109/TGRS.2007.909951 Cited on page(s) 114

B. A. Olshausen and D. J. Field. Emergence of simple-cell receptive field properties by learning a sparse code for natural images. *Nature*, 381: 607–609, 1996. DOI: 10.1038/381607a0 Cited on page(s) 27, 33

I. Olthof, C. Butson, and R. Fraser. Signature extension through space for northern landcover classification: A comparison of radiometric correction methods. *Remote Sens. Environ.*, 95 (3): 290–302, 2005. DOI: 10.1016/j.rse.2004.12.015 Cited on page(s) 81

J. E. O'Reilly, S. Maritorena, B. G. Mitchell, D. A. Siegel, K. Carder, S. A. Garver, M. Kahru, and C. McClain. Ocean color chlorophyll algorithms for SeaWiFS. *Journal of Geophysical Research*, 103 (C11): 24937–24953, Oct 1998. DOI: 10.1029/98JC02160 Cited on page(s) 108, 109, 118, 119

A. Pacheco and H. McNairn. Evaluating multispectral remote sensing and spectral unmixing analysis for crop residue mapping. *Rem. Sens. Environ.*, 114 (10): 2219–2228, 2010. DOI: 10.1016/j.rse.2010.04.024 Cited on page(s) 85

F. Pacifici, F. Del Frate, C. Solimini, and W.J. Emery. An innovative neural-net method to detect temporal changes in high-resolution optical satellite imagery. *IEEE Trans. Geosci. Rem. Sens.*, 45 (9): 2940–2952, 2007. DOI: 10.1109/TGRS.2007.902824 Cited on page(s) 69

F. Pacifici, M. Chini, and W.J. Emery. A neural network approach using multi-scale textural metrics from very high-resolution panchromatic imagery for urban land-use classification. *Remote Sens. Environ.*, 113 (6): 1276–1292, 2009a. DOI: 10.1016/j.rse.2009.02.014 Cited on page(s) 64

F. Pacifici, F. Del Frate, and W.J. Emery. Pulse coupled neural networks for detecting urban areas changes at very high resolutions. In *Urban Remote*

Sensing Event, 2009 Joint, pages 1–7, May 2009b. DOI: 10.1109/URS.2009.5137588 Cited on page(s) 69

F. Pacifici, M. Chini, C. Bignami, S. Stramondo, and W.J. Emery. Automatic damage detection using pulse-coupled neural networks for the 2009 italian earthquake. In *IEEE Geosc. Rem. Sens. Symp. (IGARSS)*, pages 1996–1999, Jul 2010. Cited on page(s) 69

E. Pagot and M. Pesaresi. Systematic study of the urban postconflict change classification performance using spectral and structural features in a support vector machine. *IEEE Journal of Selected Topics in Applied Earth Observations and Remote Sensing*, 1 (2): 120–128, Jun 2008. DOI: 10.1109/JSTARS.2008.2001154 Cited on page(s) 70

J.A. Palmason, J.A. Benediktsson, and K. Arnason. Morphological transformations and feature extraction for urban data withhigh spectral and spatial resolution. In *IEEE Geosc. Rem. Sens. Symp. (IGARSS)*, 2003. Cited on page(s) 53

M. Parente and A. Plaza. Survey of geometric and statistical unmixing algorithms for hyperspectral images. In *IEEE GRSS Workshop Hyper. Im. Sign. Proc. (WHISPERS)*, Reykjavik, Iceland, 2010. Cited on page(s) 92

B. Paskaleva, M.M. Hayat, Z. Wang, J.S. Tyo, and S. Krishna. Canonical correlation feature selection for sensors with overlapping bands: Theory and application. *IEEE Trans. Geosc. Rem. Sens.*, 46 (10): 3346–3358, Oct 2008. DOI: 10.1109/TGRS.2008.921637 Cited on page(s) 39

E. Pasolli, F. Melgani, and Y. Bazi. SVM active learning through significance space construction. *IEEE Geosci. Remote Sens. Lett.*, 8 (3): 431–435, 2011. DOI: 10.1109/LGRS.2010.2083630 Cited on page(s) 79, 80

L. Pasolli, F. Melgani, and E. Blanzieri. Gaussian process regression for estimating chlorophyll concentration in subsurface waters from remote sensing data. *IEEE Geosc. Rem. Sens. Lett.*, pages 464–468, 2010. DOI: 10.1109/LGRS.2009.2039191 Cited on page(s) 110

Y. C. Pati, R. Rezahfar, and P. Krishnaprasad. Orthogonal matching pursuit: Recursive function approximation with applications to wavelet decomposition. In *Proceedings of the 27th Annual Asilomar Conference on Signals, Systems and Computers*, Los Alamitos, CA, USA, 2003. Cited on page(s) 94

B. Penna, T. Tillo, E. Magli, and G. Olmo. Transform coding techniques for lossy hyperspectral data compression. *IEEE Trans. Geosci. Rem. Sens.*, 45 (5): 1408–1421, 2007. DOI: 10.1109/TGRS.2007.894565 Cited on page(s) 23, 26, 34

C. Persello and L. Bruzzone. A novel active learning strategy for domain adaptation in the classification of remote sensing images. In *IEEE Geosc. Rem. Sens. Symp. (IGARSS)*, Vancouver, Canada, 2011. Cited on page(s) 82

M. Pesaresi and J.A. Benediktsson. A new approach for the morphological segmentation of high-resolution satellite images. *IEEE Trans. Geosc. Rem. Sens.*, 39 (2): 309–320, 2001. DOI: 10.1109/36.905239 Cited on page(s) 50, 52, 53, 64

M. Pesaresi and I. Kannellopoulos. *Machine Vision and Advanced Image Processing in Remote Sensing*, chapter Detection of urban features using morphological based segmentation and very high resolution remotely sensed data. Springer Verlag, 1999. Cited on page(s) 53

W. R. Philipson and W. R. Hafker. Manual versus digital Landsat analysis for delineating river flooding. *Photogrammetric engineering and remote sensing*, 47: 1351–1356, 1981. Cited on page(s) 108

P. Pina and T. Barata. Classification by mathematical morphology. In *IEEE Geosc. Rem. Sens. Symp. (IGARSS)*, 2003. Cited on page(s) 53

B. Pinty and M.M. Verstraete. Extracting information surface properties from bidirectional reflectance measurements. *J. Geophys. Res.*, 96: 2865–2874, 1991. DOI: 10.1029/90JD02239 Cited on page(s) 112

A. Plaza and C.-I Chang. Impact of initialization on design of endmember extraction algorithms. *IEEE Trans. Geosc. Rem. Sens.*, 44 (11): 3397–

3407, 2006. DOI: 10.1109/TGRS.2006.879538 Cited on page(s) 92, 93

A. Plaza, P. Martínez, R. Pérez, and J. Plaza. A new method for target detection in hyperspectral imagery based on extended morphological profiles. In *IEEE Geosc. Rem. Sens. Symp. (IGARSS)*, 2003. Cited on page(s) 53

A. Plaza, P. Martinez, J. Plaza, and R. Pérez. Dimensionality reduction and classification of hyperspectral image data using sequences of extended morphological transformations. *IEEE Trans. Geosc. Rem. Sens.*, 43: 466–479, 2005. DOI: 10.1109/TGRS.2004.841417 Cited on page(s) 51, 53

A. Plaza, J. A. Benediktsson, J. Boardman, J. Brazile, L. Bruzzone, G. Camps-Valls, J. Chanussot, M. Fauvel, P. Gamba, A. Gualtieri, and J.C. Tilton. Recent advances in techniques for hyperspectral image processing. *Remote Sens. Environ.*, 113: 110–122, 2009. DOI: 10.1016/j.rse.2007.07.028 Cited on page(s) 12, 64

A. Plaza, G. Martín, J. Plaza, M. Zortea, and S. Sánchez. Recent developments in spectral unmixing and endmember extraction. In L. Bruce S. Prasad and J. Chanussot, editors, *Optical Remote Sensing-Advances in Signal Processing and Exploitation*. Springer-Verlag, 2011. Cited on page(s) 87, 93

J. Plaza, R. Pérez, A. Plaza, P. Martínez, and D. Valencia. Parallel morphological/neural processing of hyperspectral images using heterogeneous and homogeneous platforms. *Cluster Comput.*, 11: 17–32, 2008. DOI: 10.1007/s10586-007-0048-1 Cited on page(s) 64

F. C. Polcyn and D. R. Lyzenga. Landsat bathymetric mapping by multispectral processing. *Intern. Symp. Remote Sens. Environ.*, pages 1269–1276, 1979. Cited on page(s) 109

F. Prata. Land surface temperature measurement from space: AATSR algorithm theoretical basis document. Technical report, CSIRO Atmospheric Research, Jan 2002. Cited on page(s) 49

W. H. Press, B. P. Flannery, S. A. Teukolsky, and W. T. Vetterling. *Numerical Recipes. The Art of Scientific Computing, Fortran Version*.

Cambridge University Press, 1989. Cited on page(s) 113

R. Preusker, J. Fischer, A. Hünerbein, C. Brockmann, M. Zühlke, and U. Krämer. Improved MERIS cloud detection. In H. Lacoste and L. Ouwehand, editors, *Proceedings of the 2nd MERIS/(A)ATSR Workshop*, pages CD–Rom, Frascati, Italy, Nov 2008. ESA SP-666, ESA Publications Division. Cited on page(s) 49

P. Puyou-Lascassies, G. Flouzat, M. Gay, and C. Vignolles. Validation of the use of multiple linear regression as a tool for unmixing coarse spatial resolution images. *Rem. Sens. Environ.*, 49 (2): 155–166, 1994. DOI: 10.1016/0034-4257(94)90052-3 Cited on page(s) 85

J. Qi, A. Chehbouni, A. R. Huete, Y. H. Kerr, and S. Sorooshian. A modified soil adjusted vegetation index. *Rem. Sens. Environ.*, 48: 119–126, 1994. DOI: 10.1016/0034-4257(94)90134-1 Cited on page(s) 106

Sarker. L. R. and J. E. Nichol. Improved forest biomass estimates using ALOS AVNIR-2 texture indices. *Rem. Sens. Environ.*, 115 (4): 968–977, 2011. DOI: 10.1016/j.rse.2010.11.010 Cited on page(s) 106

R. J. Radke, S. Andra, O. Al-Kofahi, and B. Roysam. Image Change Detection Algorithms: A Systematic Survey. *IEEE Trans. Image Process.*, 14 (3): 294–307, 2005. DOI: 10.1109/TIP.2004.838698 Cited on page(s) 67

S. Rajan, J. Ghosh, and M. Crawford. An active learning approach to knowledge transfer for hyperspectral data analysis. In *IEEE Geosc. Rem. Sens. Symp. (IGARSS)*, Denver, USA, 2006a. Cited on page(s) 82

S. Rajan, J. Ghosh, and M. Crawford. Exploiting class hierarchy for knowledge transfer in hyperspectral data. *IEEE Trans. Geosci. Rem. Sens.*, 44 (11): 3408–3417, 2006b. DOI: 10.1109/TGRS.2006.878442 Cited on page(s) 81

S. Rajan, J. Ghosh, and M. Crawford. An active learning approach to hyperspectral data classification. *IEEE Trans. Geosci. Rem. Sens.*, 46 (4): 1231–1242, 2008. DOI: 10.1109/TGRS.2007.910220 Cited on page(s) 78, 80

C. E. Rasmussen and C. K. I. Williams. *Gaussian Processes for Machine Learning*. The MIT Press, New York, 2006. Cited on page(s) 107, 116

M. Rast, J.L. Bézy, and S. Bruzzi. The ESA Medium Resolution Imaging Spectrometer MERIS: a review of the instrument and its mission. *Int. Journal of Remote Sensing*, 20 (9): 1681–1702, Jun 1999. DOI: 10.1080/014311699212416 Cited on page(s) 12, 41

F. Ratle, G. Camps-Valls, and J. Weston. Semisupervised neural networks for efficient hyperspectral image classification. *IEEE Trans. Geosci. Rem. Sens.*, 48 (5): 2271–2282, May 2010. DOI: 10.1109/TGRS.2009.2037898 Cited on page(s) 76

I. S. Reed and X. Yu. Adaptive multiple-band CFAR detection of an optical pattern with unknown spectral distribution. *IEEE Trans. Acoustics, Speech and Signal Process.*, 38 (10): 1760–1770, 1990. DOI: 10.1109/29.60107 Cited on page(s) 73, 74

G. Rellier, X. Descombes, F. Falzon, and J. Zerubia. Texture feature analysis using a Gauss-Markov model in hyperspectral image classification. *IEEE Trans. Geosc. Rem. Sens.*, 42 (7): 1543–1551, Jul 2004. DOI: 10.1109/TGRS.2004.830170 Cited on page(s) 50, 51

N. Renard and S. Bourennane. Dimensionality reduction based on tensor modeling for classification methods. *IEEE Trans. Geosci. Rem. Sens.*, 47 (4): 1123–1131, Apr 2009. DOI: 10.1109/TGRS.2008.2008903 Cited on page(s) 55

J.-M. Renders and S.P. Flasse. Hybrid methods using genetic algorithms for global optimization. *IEEE Trans. Syst. Man Cyb.*, 26: 243–258, 1996. DOI: 10.1109/3477.485836 Cited on page(s) 112

J. Rhee, J. Im, G. J. Carbone, and J. R. Jensen. Delineation of climate regions using in-situ and remotely-sensed data for the Carolinas. *Remote Sens. Environ.*, 112 (6): 3099–3111, 2008. DOI: 10.1016/j.rse.2008.03.001 Cited on page(s) 65

J. A. Richards. Analysis of remotely sensed data: the formative decades and the future. *IEEE Trans. Geosci. Rem. Sens.*, 43 (3): 422–432, 2005. DOI:

10.1109/TGRS.2004.837326 Cited on page(s) 19, 20

J. A. Richards and Xiuping Jia. *Remote Sensing Digital Image Analysis. An Introduction.* Springer-Verlag, Berlin, Heidelberg, Germany, 3rd edition, 1999. Cited on page(s) 2, 3, 11

D.A Roberts, M Gardner, R Church, S Ustin, G Scheer, and R.O. Green. Mapping chaparral in the Santa Monica Mountains using multiple endmember spectral mixture models. *Rem. Sens. Environ.*, 65 (3): 267–279, 1998. DOI: 10.1016/S0034-4257(98)00037-6 Cited on page(s) 84, 93

F.C. Robey, D.R. Fuhrmann, E.J. Kelly, and R. Nitzberg. A CFAR adaptive matched filter detector. *IEEE Trans. Aerosp. Elect. Syst.*, 28 (1): 208–216, Jan 1992. DOI: 10.1109/7.135446 Cited on page(s) 74

C. D. Rodgers. *Inverse Methods for Atmospheric Sounding: Theory and Practice.* World Scientific Publishing Co. Ltd., 2000. Cited on page(s) 4, 101, 110, 116

D. M. Rogge, B. Rivard, J. Zhang, and J Feng. Iterative spectral unmixing for optimizing per-pixel endmember sets. *IEEE Trans. Geosc. Rem. Sens.*, 44 (12): 3725–3736, 2006. DOI: 10.1109/TGRS.2006.881123 Cited on page(s) 94

G. Rondeaux, M. Steven, and F. Baret. Optimization of soil-adjusted vegetation indices. *Rem. Sens. Environ.*, 55 (2): 95–107, 1996. DOI: 10.1016/0034-4257(95)00186-7 Cited on page(s) 106

J.W. Rouse, R.H. Haas, J.A. Schell, and D.W. Deering. Monitoring vegetation systems in the great plains with ERTS. In *Third ERTS Symposium, NASA SP-351 I*, pages 309–317, 1973. Cited on page(s) 48, 105

S.T. Roweis and L.K. Saul. Nonlinear dimensionality reduction by locally linear embedding. *Science*, 290 (5500): 2323–2326, 2000. DOI: 10.1126/science.290.5500.2323 Cited on page(s) 43

D.L. Ruderman and W. Bialek. Statistics of natural images: Scaling in the woods. *Physical Review Letters*, 73 (6): 814–817, 1994. DOI: 10.1103/PhysRevLett.73.814 Cited on page(s) 20

J.A. Saghri, A.G. Tescher, and J.T. Reagan. Practical transform coding of multispectral imagery. *IEEE Sig. Proc. Mag.*, 12 (1): 32–43, 1995. DOI: 10.1109/79.363506 Cited on page(s) 29

V.V. Salomonson, W.L. Barnes, P.W. Maymon, H.E. Montgomery, and H. Ostrow. MODIS: advanced facility instrument for studies of the Earth as a system. *IEEE Trans. Geosc. Rem. Sens.*, 27 (2): 145–153, Mar 1989. DOI: 10.1109/36.20292 Cited on page(s) 12

A. Sarkar, M.K. Biswas, B. Kartikeyan, V. Kumar, K.L. Majumder, and D.K. Pal. A MRF model-based segmentation approach to classification for multispectral imagery. *IEEE Trans. Geosci. Rem. Sens.*, 40 (5): 1102–1113, May 2002. DOI: 10.1109/TGRS.2002.1010897 Cited on page(s) 64

L.K. Saul, K.Q. Weinberger, J.H. Ham, F. Sha, and D.D. Lee. *Spectral methods for dimensionality reduction*, chapter 16, pages 293–308. MIT Press, 2006. Cited on page(s) 43

M.E. Schaepman, R.O. Green, S. Ungar, B. Curtiss, J. Boardman, A. Plaza, B.-C. Gao, R. Kokaly, J.R. Miller, S. Jacquemoud, E. Ben-Dor, R. Clark, C. Davis, J. Dozier, D. Goodenough, D.A. Roberts, and A. Goetz. The Future of Imaging Spectroscopy – Prospective Technologies and Applications. In *IEEE Geosc. Rem. Sens. Symp. (IGARSS)*, pages 2005–2009, Denver, CO, USA, Jul 2006. Cited on page(s) 12, 14

M.E. Schaepman, S.L. Ustin, A.J. Plaza, T.H. Painter, J. Verrelst, and S. Liang. Earth system science related imaging spectroscopy-An assessment. *Rem. Sens. Environ.*, 113 (1): S123–S137, 2009. DOI: 10.1016/j.rse.2009.03.001 Cited on page(s) 48, 105

L.L. Scharf and B. Friedlander. Matched subspace detectors. *Signal Processing, IEEE Transactions on*, 42 (8): 2146–2157, August 1994. Cited on page(s) 74

A. Schaum and A. Stocker. Hyperspectral change detection and supervised matched filtering based on covariance equalization. In *Proc. SPIE 5425*, pages 77–90, 1994. Cited on page(s) 73

A. Schaum and A. Stocker. Spectrally selective target detection. In *Proceedings of the International Symposium on Spectral Sensing Research*, 1997a. Cited on page(s) 73

A. Schaum and A. Stocker. Long-interval chronochrome target detection. In *Proceedings of the International Symposium on Spectral Sensing Research*, 1997b. Cited on page(s) 73

B. Schölkopf and A. Smola. *Learning with Kernels – Support Vector Machines, Regularization, Optimization and Beyond*. MIT Press Series, Cambridge, MA, USA, 2002. Cited on page(s) 43

R. A. Schowengerdt. *Remote Sensing: Models and Methods for Image Processing*. Academic, San Diego, USA, 1997. Cited on page(s) 97

G. Schwarz. Estimating the dimension of a model. *Annals of Statistics*, 6: 461–464, 1978. DOI: 10.1214/aos/1176344136 Cited on page(s) 89

S.B. Serpico and L. Bruzzone. A new search algorithm for feature selection in hyperspectral remote sensing images. *IEEE Trans. Geosc. Rem. Sens.*, 39 (7): 1360–1367, Jul 2001. DOI: 10.1109/36.934069 Cited on page(s) 39, 40

S.B. Serpico and G. Moser. Extraction of spectral channels from hyperspectral images for classification purposes. *IEEE Trans. Geosc. Rem. Sens.*, 45 (2): 484–495, Feb 2007. DOI: 10.1109/TGRS.2006.886177 Cited on page(s) 39

J. Serra. *Image analysis and Mathematical Morphology*. Academic, 1982. Cited on page(s) 51

J. Settle. On the effect of variable endmember spectra in the linear mixture model. *IEEE Trans. Geosc. Rem. Sens.*, 44 (2): 389–396, 2006. DOI: 10.1109/TGRS.2005.860983 Cited on page(s) 93

V.P. Shah, N.H. Younan, S.S. Durbha, and R.L. King. Feature identification via a combined ICA-wavelet method for image information mining. *IEEE Geosc. Rem. Sens. Lett.*, 7 (1): 18–22, Jan 2010. Cited on page(s) 50, 55

P. Shanmugam, B. Sundarabalan, Ahn Yu-Hwan, and Ryu Joo-Hyung. A new inversion model to retrieve the particulate backscattering in coastal/ocean waters. *IEEE Trans. Geosc. Rem. Sens.*, 49 (6): 2463–2475, Jun 2011. DOI: 10.1109/TGRS.2010.2103947 Cited on page(s) 109

G. Sharma, M. J. Vrhel, and H.J. Trusell. Color imaging for multimedia. *Proc. IEEE*, 86 (6): 1088–1108, 1998. DOI: 10.1109/5.687831 Cited on page(s) 19

G. Shaw and D. Manolakis. Signal processing for hyperspectral image exploitation. *IEEE Signal Proc. Magazine*, 50: 12–16, Jan 2002. DOI: 10.1109/79.974715 Cited on page(s) 2, 83

J. Shawe-Taylor and N. Cristianini. *Kernel Methods for Pattern Analysis*. Cambridge University Press, Cambridge, MA, USA, 2004. Cited on page(s) 43

M. Shoshany and T. Svoray. Multidate adaptive unmixing and its application to analysis of ecosystem transitions along a climatic gradient. *Rem. Sens. Environ.*, 82 (1): 5–20, 2002. DOI: 10.1016/S0034-4257(01)00346-7 Cited on page(s) 85

D. Siméoni, C. Singer, and G. Chalon. Infrared atmospheric sounding interferometer. *Acta Astronautica*, 40: 113–118, 1997. DOI: 10.1016/S0094-5765(97)00098-2 Cited on page(s) 21, 22, 120

K.K. Simhadri, S.S. Iyengar, R.J. Holyer, M. Lybanon, and Jr Zachary, J.M. Wavelet-based feature extraction from oceanographic images. *IEEE Trans. Geosci. Rem. Sens.*, 36 (3): 767–778, May 1998. DOI: 10.1109/36.673670 Cited on page(s) 54

E.P. Simoncelli. Statistical models for images: Compression, restoration and synthesis. In *31st Asilomar Conference on Signals, Syst. and Comp., Pacific Grove, CA*. IEEE, 1997. Cited on page(s) 21, 22, 27, 33

R. B. Singer and T. B. McCord. Mars: Large scale mixing of bright and dark surface materials and implications for analysis of spectral reflectance. In *Proc. of the 10th Lunar and Planetary Sci. Conf*, pages 1835–1848, 1979. Cited on page(s) 86

A. Singh. Digital change detection techniques using remotely-sensed data. *Int. J. Rem. Sens.*, 10 (6): 989–1003, 1989. DOI: 10.1080/01431168908903939 Cited on page(s) 67, 68, 69

Barun Singh, William T. Freeman, and David H. Brainard. Exploiting spatial and spectral image regularities for color constancy. In *Proc. 3rd Int'l Workshop on Statis. and Compu. Theories of Vision*, Nice France, October 2003. Cited on page(s) 20, 25

W.H. Slade, H.W. Ressom, M.T. Musavi, and R.L. Miller. Inversion of ocean color observations using particle swarm optimization. *IEEE Trans. Geosc. Rem. Sens.*, 42 (9): 1915–1923, Sep 2004. DOI: 10.1109/TGRS.2004.833389 Cited on page(s) 113

C. Small. High spatial resolution spectral mixture analysis of urban reflectance. *Rem. Sens. Environ.*, 88 (1-2): 170–186, 2003. DOI: 10.1016/j.rse.2003.04.008 Cited on page(s) 86

J.A. Smith. LAI inversion using backpropagation neural network trained with multiple scattering model. *IEEE Trans. Geosc. Rem. Sens.*, 31 (5): 1102–1106, 1993. DOI: 10.1109/36.263783 Cited on page(s) 115

M.O. Smith, S.L. Ustin, J.B. Adams, and A.R. Gillespie. Vegetation in deserts: I. a regional measure of abundance from multispectral images. *Rem. Sens. Environ.*, 31 (1): 1–26, 1990. DOI: 10.1016/0034-4257(90)90074-V Cited on page(s) 84

A. J. Smola and B. Schölkopf. A tutorial on support vector regression. *Statistics and Computing*, 14: 199–222, 2004. DOI: 10.1023/B:STCO.0000035301.49549.88 Cited on page(s) 107, 116, 117

L.K. Soh and C. Tsatsoulis. Texture analysis of SAR sea ice imagery using gray level co-occurrence matrices. *IEEE Trans. Geosci. Rem. Sens.*, 37 (2): 780–795, Mar 1999. DOI: 10.1109/36.752194 Cited on page(s) 51

Y. Sohn and R.M. McCoy. Mapping desert shrub rangeland using spectral unmixing and modeling spectral mixtures with TM data. *Photogrammetric Engineering and Remote Sensing*, 63 (6): 707–716, 1997. Cited on page(s) 84

P. Soille. *Morphological image analysis*. Springer-Verlag, Berlin-Heidelberg, 2004. Cited on page(s) 51

P. Soille and M. Pesaresi. Advances in mathematical morphology applied to geoscience and remote sensing. *IEEE Trans. Geosc. Rem. Sens.*, 40 (9): 2042–2055, 2002. DOI: 10.1109/TGRS.2002.804618 Cited on page(s) 53

S. Solbo and T. Eltoft. Homomorphic wavelet-based statistical despeckling of SAR images. *IEEE Trans. Geosc. Rem. Sens.*, 42 (4): 711–721, Apr 2004. DOI: 10.1109/TGRS.2003.821885 Cited on page(s) 21

C. Song. Spectral mixture analysis for subpixel vegetation fractions in the urban environment: How to incorporate endmember variability? *Rem. Sens. Environ.*, 95 (2): 248–263, 2005. DOI: 10.1016/j.rse.2005.01.002 Cited on page(s) 93

S. Stagakis, N. Markos, O. Sykioti, and A. Kyparissis. Monitoring canopy biophysical and biochemical parameters in ecosystem scale using satellite hyperspectral imagery: An application a phlomis fruticosa mediterranean ecosystem using multiangular CHRIS/PROBA observations. *Rem. Sens. Environ.*, 114 (5): 977–994, 2010. DOI: 10.1016/j.rse.2009.12.006 Cited on page(s) 105

S.R. Sternberg. Grayscale morphology. *Computer Vision Graphics and Image Processing*, 35: 333–355, 1986. DOI: 10.1016/0734-189X(86)90004-6 Cited on page(s) 51, 52, 53

W. Stiles and G. Wyszecki. *Color Science: Concepts and Methods, Quantitative Data and Formulae*. John Wiley and sons, New York, 1982. Cited on page(s) 22, 24

M.P. Stoll, C. Buschmann, A. Court, T. Laurila, J. Moreno, and I. Moya. The FLEX-Fluorescence Explorer mission project: motivations and present status of preparatory activities. In *IEEE Geosc. Rem. Sens. Symp. (IGARSS)*, volume 1, pages 585–587, 2003. Cited on page(s) 14

W.D. Stromberg and T.G. Farr. A Fourier-based textural feature extraction procedure. *IEEE Trans. Geosci. Rem. Sens.*, GE-24 (5): 722–731, Sep 1986. DOI: 10.1109/TGRS.1986.289620 Cited on page(s) 54

T. Stuffler, C. Kaufmann, S. Hofer, K. P. Förster, G. Schreier, A. Mueller, A. Eckardt, H. Bach, B. Penné, U. Benz, and R. Haydn. The EnMAP hyperspectral imager – An advanced optical payload for future applications in Earth observation programmes. *Acta Astronautica*, 61 (1-6): 115–120, Jun 2007. DOI: 10.1016/j.actaastro.2007.01.033 Cited on page(s) 14

S. Tadjudin and D. A. Landgrebe. Robust parameter estimation for mixture model. *IEEE Trans. Geosci. Rem. Sens.*, 38 (1): 439–445, 2000. DOI: 10.1109/36.823939 Cited on page(s) 76

X. Tang and W.A. Pearlman. *Three-dimensional wavelet-based compression of hyperspectral images*, pages 273–308. Springer, MA: Kluwer, 2005. Cited on page(s) 34

Y. Tarabalka, J.A. Benediktsson, and J. Chanussot. Spectral-spatial classification of hyperspectral imagery based on partitional clustering techniques. *IEEE Trans. Geosci. Rem. Sens.*, 47 (8): 2973–2987, 2009. DOI: 10.1109/TGRS.2009.2016214 Cited on page(s) 64

Y. Tarabalka, J. Chanussot, and J.A. Benediktsson. Segmentation and classification of hyperspectral images using minimum spanning forest grown from automatically selected markers. *Systems, Man, and Cybernetics, Part B: Cybernetics, IEEE Transactions on*, 40 (5): 1267–1279, 2010. Cited on page(s) 65

J.B. Tenenbaum, V. de Silva, and J.C. Langford. A global geometric framework for nonlinear dimensionality reduction. *Science*, 290 (5500): 2319–2323, 2000. DOI: 10.1126/science.290.5500.2319 Cited on page(s) 43

J. Theiler. Sensitivity of anomalous change detection to small misregistration errors. In *Proc. SPIE 6966*, 2008a. Cited on page(s) 73

J. Theiler. Quantitative comparison of quadratic covariance-based anomalous change detectors. *Appl. Opt.*, 47: F12–F26, 2008b. DOI: 10.1364/AO.47.000F12 Cited on page(s) 73

J. Theiler and S. Perkins. Proposed framework for anomalous change detection. In *Proc. ICML Workshop Mach. Learn. Algorithms Surveill. Event Detect.*, pages 7–14, 2006. Cited on page(s) 73

J. Theiler, G. Cao, L. R. Bachega, and C. A. Bouman. Sparse matrix transform for hyperspectral image processing. *IEEE J. Sel. Topics Signal Proc.*, 5 (3): 424–437, 2011. DOI: 10.1109/JSTSP.2010.2103924 Cited on page(s) 73

G. Thuillier, M. Hersé, D. Labs,T. Foujols, W. Peetermans, D. Gillotay, P. C. Simon, and H. Mandel. The solar spectral irradiance from 200 to 2400 nm as measured by the SOLSPEC spectrometer from the ATLAS and EURECA missions. *Solar Physics*, 214: 1–22, 2003. DOI: 10.1023/A:1024048429145 Cited on page(s) 6

Y.M. Timofeyev, A.V. Polyakov, H.M. Steele, and M.J. Newchurch. Optimal eigenanalysis for the treatment of aerosols in the retrieval of atmospheric composition from transmission measurements. *Applied Optics*, 42 (15): 2635–2646, 2003. DOI: 10.1364/AO.42.002635 Cited on page(s) 23

M. E. Tipping. Sparse Bayesian Learning and the Relevance Vector Machine. *Journal of Machine Learning Research*, 1: 211–244, 2001. Cited on page(s) 110

S. Tompkins, J.F Mustard, C.M Pieters, and D.W. Forsyth. Optimization of endmembers for spectral mixture analysis. *Rem. Sens. Environ.*, 59 (3): 472–489, 1997. DOI: 10.1016/S0034-4257(96)00122-8 Cited on page(s) 92

P.A. Townsend, J.R. Foster, R.A. Jr. Chastain, and W.S. Currie. Application of imaging spectroscopy to mapping canopy nitrogen in the forests of the central Appalachian Mountains using Hyperion and AVIRIS. *IEEE Trans. Geosc. Rem. Sens.*, 41 (6): 1347–1354, June 2003. DOI: 10.1109/TGRS.2003.813205 Cited on page(s) 107

T.M. Tu. Unsupervised signature extraction and separation in hyperspectral images: A noise-adjusted fast independent component analysis approach. In *Opt. Eng./SPIE*, volume 39(4), pages 897–906, 2000. DOI: 10.1117/1.602461 Cited on page(s) 92

C.J. Tucker. Red and photographic infrared linear combinations for monitoring vegetation. *Rem. Sens. Environ.*, 8 (2): 127–150, 1979. DOI: 10.1016/0034-4257(79)90013-0 Cited on page(s) 48

D. Tuia and G. Camps-Valls. Semi-supervised remote sensing image classification with cluster kernels. *IEEE Geosci. Remote Sens. Lett.*, 6 (1): 224–228, 2009. DOI: 10.1109/LGRS.2008.2010275 Cited on page(s) 76

D. Tuia, F. Pacifici, M. Kanevski, and W.J. Emery. Classification of very high spatial resolution imagery using mathematical morphology and support vector machines. *IEEE Trans. Geosci. Rem. Sens.*, 47 (11): 3866–3879, Nov 2009a. Cited on page(s) 53, 64

D. Tuia, F. Ratle, F. Pacifici, M. Kanevski, and W.J. Emery. Active learning methods for remote sensing image classification. *IEEE Trans. Geosci. Rem. Sens.*, 47 (7): 2218–2232, 2009b. DOI: 10.1109/TGRS.2008.2010404 Cited on page(s) 78, 79, 80

D. Tuia, G. Camps-Valls, G. Matasci, and M. Kanevski. Learning relevant image features with multiple kernel classification. *IEEE Trans. Geosci. Rem. Sens.*, 48 (10): 3780–3791, 2010a. DOI: 10.1109/TGRS.2010.2049496 Cited on page(s) 64

D. Tuia, J. Muñoz-Marí, M. Kanevski, and G. Camps-Valls. Cluster-based active learning for compact image classification. In *IEEE Geosc. Rem. Sens. Symp. (IGARSS)*, Hawaii, USA, 2010b. Cited on page(s) 79

D. Tuia, F. Ratle, A. Pozdnoukhov, and G. Camps-Valls. Multi-source composite kernels for urban image classification. *IEEE Geosci. Remote*

Sens. Lett., Special Issue ESA_EUSC, 7 (1): 88–92, 2010c. Cited on page(s) 64

D. Tuia, E. Pasolli, and W. J. Emery. Using active learning to adapt remote sensing image classifiers. *Rem. Sens. Environ.*, 115 (9): 2232–2242, 2011a. DOI: 10.1016/j.rse.2011.04.022 Cited on page(s) 79, 82

D. Tuia, J. Verrelst, L. Alonso, F. Pérez-Cruz, and G. Camps-Valls. Multioutput support vector regression for remote sensing biophysical parameter estimation. *IEEE Geosc. Rem. Sens. Lett.*, 8 (4): 804–808, 2011b. DOI: 10.1109/LGRS.2011.2109934 Cited on page(s) 107, 116, 117

D. Tuia, M. Volpi, L. Copa, M. Kanevski, and J. Muñoz-Marí. A survey of active learning algorithms for supervised remote sensing image classifications:. *IEEE J. Sel. Topics Signal Proc.*, 5 (3): 606–617, 2011c. Cited on page(s) 80

M. Tyagi, F. Bovolo, A. Mehra, S. Chaudhuri, and L. Bruzzone. A context-sensitive clustering technique based on graph-cut initialization and expectation-maximization algorithm. *IEEE Geosci. Remote Sens. Lett.*, 5 (1): 15–21, 2008. DOI: 10.1109/LGRS.2007.905119 Cited on page(s) 64

S.G. Ungar, J.S. Pearlman, J.A. Mendenhall, and D. Reuter. Overview of the Earth Observing One (EO-1) mission. *IEEE Trans. Geosc. Rem. Sens.*, 41 (6): 1149–1159, Jun 2003. DOI: 10.1109/TGRS.2003.815999 Cited on page(s) 14

S. Ustin. *Remote Sensing for Natural Resource Management and Environmental Monitoring. Manual of Remote Sensing, Volume 4*. John Wiley & Sons, New York, 2004. Cited on page(s) 2, 3

B.G. Vasudevan, B.S. Gohil, and V.K. Agarwal. Backpropagation neural-network-based retrieval of atmospheric water vapor and cloud liquid water from IRS-P4 MSMR. *IEEE Trans. Geosc. Rem. Sens.*, 42 (5): 985–990, may 2004. DOI: 10.1109/TGRS.2004.825580 Cited on page(s) 115

A. Verger, F. Baret, and F. Camacho. Optimal modalities for radiative transfer-neural network estimation of canopy biophysical characteristics:

Evaluation over an agricultural area with CHRIS/PROBA observations. *Rem. Sens. Environ.*, 115 (2): 415–426, 2011. DOI: 10.1016/j.rse.2010.09.012 Cited on page(s) 115

W. Verhoef. Light scattering by leaf layers with application to canopy reflectance modeling: the SAIL model. *Rem. Sens. Environ.*, 16: 125–141, 1984. DOI: 10.1016/0034-4257(84)90057-9 Cited on page(s) 112

W. Verhoef and H. Bach. Simulation of hyperspectral and directional radiance images using coupled biophysical and atmospheric radiative transfer models. *Rem. Sens. Environ.*, 87: 23–41, 2003. DOI: 10.1016/S0034-4257(03)00143-3 Cited on page(s) 111

J. Verrelst, M.E. Schaepman, B. Koetz, and M. Kneubühler. Angular sensitivity analysis of vegetation indices derived from CHRIS/PROBA data. *Rem. Sens. Environ.*, 112 (5): 2341–2353, 2008. DOI: 10.1016/j.rse.2007.11.001 Cited on page(s) 105

J. Verrelst, L. Alonso, G. Camps-Valls, J. Delegido, and J. Moreno. Retrieval of canopy parameters using Gaussian processes techniques. *IEEE Trans. Geosc. Rem. Sens.*, 49, 2011. DOI: 10.1109/TGRS.2011.2168962 Cited on page(s) 107, 116, 117

M. M. Verstraete and B. Pinty. Potential and limitations of information extraction the terrestrial biosphere from satellite remote sensing. *Rem. Sens. Environ.*, 58: 201–214, 1996. DOI: 10.1016/S0034-4257(96)00069-7 Cited on page(s) 103

M. Volpi, D. Tuia, M. Kanevski, and G. Camps-Valls. Unsupervised change detection by kernel clustering. In *Proceedings of the SPIE Remote Sensing Conference*, 2010. Cited on page(s) 69

M. Volpi, D. Tuia, F. Bovolo, M. Kanevski, and L. Bruzzone. Supervised change detection in VHR images using contextual information and support vector machines. *International Journal of Applied Earth Observation and Geoinformation*, 2011a. Cited on page(s) 70

M. Volpi, D. Tuia, and M. Kanevski. Cluster-based active learning for remote sensing image classification. *IEEE Trans. Geosci. Rem. Sens.*,

2011b. Cited on page(s) 79, 80

T. Wachtler, T. Lee, and T. J. Sejnowski. Chromatic structure of natural scenes. *JOSA A*, 18 (1): 65–77, 2001. DOI: 10.1364/JOSAA.18.000065 Cited on page(s) 26

G.K. Wallace. The JPEG still picture compression standard. *Communications of the ACM*, 34 (4): 31–43, 1991. DOI: 10.1145/103085.103089 Cited on page(s) 29

B-C. Wang. *Digital Signal Processing Techniques and Applications in Radar Image Processing*. John Wiley & Sons, New York, 2008. Cited on page(s) 2

D. Wang, D.C. He, and D. Morin. Classification of remotely sensed images using mathematical morphology. In *IEEE Geosc. Rem. Sens. Symp. (IGARSS)*, 1994. Cited on page(s) 53

D. Wang, D. S. Yeung, and E.C.C. Tsang. Structured one-class classification. *Systems, Man, and Cybernetics, Part B: Cybernetics, IEEE Transactions on*, 36 (6): 1283–1295, Dec 2006. Cited on page(s) 74

J. Wang and C.-I. Chang. Independent component analysis-based dimensionality reduction with applications in hyperspectral image analysis. *IEEE Trans. Geosc. Rem. Sens.*, 44 (6): 1586–1600, 2006a. DOI: 10.1109/TGRS.2005.863297 Cited on page(s) 89

J. Wang and C.I. Chang. Independent component analysis-based dimensionality reduction with applications in hyperspectral image analysis. *IEEE Trans. Geosci. Rem. Sens.*, 44 (6): 1586–1600, Jun 2006b. DOI: 10.1109/TGRS.2005.863297 Cited on page(s) 42

Y Wang and Y. Jin. A genetic algorithm to simultaneously retrieve land surface roughness and soil moisture. *Journal of Remote Sensing*, 4: 90–94, 2000. Cited on page(s) 113

D. Q. Wark. On indirect temperature soundings of the stratosphere from satellites. *J. Geophys. Res.*, 6: 77, 1961. Cited on page(s) 110

B. Waske and J.A. Benediktsson. Fusion of support vector machines for classification of multisensor data. *IEEE Trans. Geosci. Rem. Sens.*, 45 (12): 3858–3866, 2007. DOI: 10.1109/TGRS.2007.898446 Cited on page(s) 64

B. Waske, S. van der Linden, J. A. Benediktsson, A. Rabe, and P. Hostert. Sensitivity of support vector machines to random feature selection in classification of hyperspectral data. *IEEE Trans. Geosci. Rem. Sens.*, 48 (10): 3747–3762, 2010. Cited on page(s) 64

M. Weiss and F. Baret. Evaluation of canopy biophysical variable retrieval performances from the accumulation of large swath satellite data. *Remote Sens. Environ*, 70: 293–306, 19991. DOI: 10.1016/S0034-4257(99)00045-0 Cited on page(s) 115

M. Weiss, F. Baret, R. Myneni, A. Pragnere, and Y. Knyazikhin. Investigation of a model inversion technique to estimate canopy biophysical variables from spectral and directional reflectance data. *Agronomie*, 20 (1): 3–22, 2000. DOI: 10.1051/agro:2000105 Cited on page(s) 114

Jason Weston, Andre Elisseeff, Berhard Scholkopf, Mike Tipping, and Pack Kaelbling. Use of the zero-norm with linear models and kernel methods. *Journal of Machine Learning Research*, 3: 1439–1461, 2003. Cited on page(s) 40

J.S. Weszka, C.R. Dyer, and A. Rosenfeld. A comparative study of texture measures for terrain classification. *Systems, Man and Cybernetics, IEEE Transactions on*, SMC-6 (4): 269–285, Apr 1976. Cited on page(s) 54

M. Wettle, Brando. V. E., and A. G. Dekker. A methodology for retrieval of environmental noise equivalent spectra applied to four Hyperion scenes of the same tropical coral reef. *Rem. Sens. Environ.*, 93 (1-2): 188–197, Oct 2004. DOI: 10.1016/j.rse.2004.07.014 Cited on page(s) 21

R. H. Whittaker and P. L. Marks. Methods of assessing terrestrial productivity. *Primary Productivity of the Biosphere*, pages 55–118, 1975. Cited on page(s) 104

M. E. Winter. N-FINDR: an algorithm for fast autonomous spectral end-member determination in hyperspectral data. In *Proc. of the SPIE conference on Imaging Spectrometry V*, pages 266–275, 1999. Cited on page(s) 94

H. Wold. *Multivariate Analysis*. Academic Press, 1966. Cited on page(s) 42

J.E. Womack and J.R. Cruz. Seismic data filtering using a Gabor representation. *IEEE Trans. Geosci. Rem. Sens.*, 32 (2): 467–472, Mar 1994. DOI: 10.1109/36.295061 Cited on page(s) 50

C. Wu. Normalized spectral mixture analysis for monitoring urban composition using etm+ imagery. *Rem. Sens. Environ.*, 93 (4): 480–492, 2004. DOI: 10.1016/j.rse.2004.08.003 Cited on page(s) 85

X. Xie, W. Timothy Liu, and B. Tang. Spacebased estimation of moisture transport in marine atmosphere using support vector regression. *Rem. Sens. Environ.*, 112 (4): 1846–1855, 2008. DOI: 10.1016/j.rse.2007.09.003 Cited on page(s) 111

M. Xu and P.K. Varshney. A subspace method for Fourier-based image registration. *IEEE Geosc. Rem. Sens. Lett.*, 6 (3): 491–494, Jul 2009. DOI: 10.1109/LGRS.2009.2018705 Cited on page(s) 54

G. Yamamoto. Numerical method for estimating the stratospheric temperature distribution from satellite measurements in the CO_2 band. *J. Met.*, 18: 581, 1961. DOI: 10.1175/1520-0469(1961)018%3C0581:NMFETS%3E2.0.CO;2 Cited on page(s) 110

F. Yang, M.A. White, A.R. Michaelis, K. Ichii, H. Hashimoto, P. Votava, A-Xing Zhu, and R.R. Nemani. Prediction of continental-scale evapotranspiration by combining modis and ameriflux data through support vector machine. *IEEE Trans. Geosc. Rem. Sens.*, 44 (11): 3452–3461, nov. 2006. DOI: 10.1109/TGRS.2006.876297 Cited on page(s) 111

C. Yangchi, M. Crawford, and J. Ghosh. Applying nonlinear manifold learning to hyperspectral data for land cover classification. In *Proc. of the*

IEEE Int. Geoscience and Remote Sensing Symp, volume 6, pages 4311–4314, 2005. Cited on page(s) 89

P.J. Zarco-Tejada, J.R. Miller, T.L. Noland, G.H. Mohammed, and P.H Sampson. Scaling-up and model inversion methods with narrowband optical indices for chlorophyll content estimation in closed forest canopies with hyperspectral data. *IEEE Trans. Geosc. Rem. Sens.*, 39 (7): 1491–1507, 2001. DOI: 10.1109/36.934080 Cited on page(s) 113

A. Zare and P. Gader. Sparsity promoting iterated constrained endmember detection in hyperspectral imagery. *IEEE Geosc. Rem. Sens. Lett.*, 4 (3): 446–450, Jul 2007. DOI: 10.1109/LGRS.2007.895727 Cited on page(s) 94

H. Zhan. *Application of Support Vector Machines in Inverse Problems in Ocean Color Remote Sensing*, chapter 18, pages 387–397. Studies in Fuzziness and Soft Computing. Springer-Verlag, Berlin, 1st edition, 2005. Cited on page(s) 110

H. Zhan, Z. Lee, P. Shi, C. Chen, and K.L. Carder. Retrieval of water optical properties for optically deep waters using genetic algorithms. *IEEE Trans. Geosc. Rem. Sens.*, 41 (5): 1123–1128, May 2003. DOI: 10.1109/TGRS.2003.813554 Cited on page(s) 113

J. Zhang, B. Rivard, and A. Sánchez-Azofeifa. Derivative spectral unmixing of hyperspectral data applied to mixtures of lichen and rock. *IEEE Trans. Geosc. Rem. Sens.*, 42 (9): 1934–1940, Sep 2004. DOI: 10.1109/TGRS.2004.832239 Cited on page(s) 85

J. Zhang, B. Rivard, and A. Sánchez-Azofeifa. Spectral unmixing of normalized reflectance data for the deconvolution of lichen and rock mixtures. *Rem. Sens. Environ.*, 95 (1): 57–66, 2005a. DOI: 10.1016/j.rse.2004.11.019 Cited on page(s) 85

L. Zhang, Y. Zhong, B. Huang, J. Gong, and P. Li. Dimensionality reduction based on clonal selection for hyperspectral imagery. *IEEE Trans. Geosci. Rem. Sens.*, 45 (12): 4172–4186, Dec 2007. DOI: 10.1109/TGRS.2007.905311 Cited on page(s) 40

X. Zhang and C.H. Chen. New independent component analysis method using higher order statistics with application to remote sensing images. *Opt. Eng.*, 41 (7): 1717–1728, 2002. DOI: 10.1117/1.1482722 Cited on page(s) 42, 109

Y.J. Zhang, C.J. Zhao, L.Y. Liu, J.H. Wang, and R.C. Wang. Chlorophyll fluorescence detected passively by difference reflectance spectra of wheat (*triticum aestivum l.*) leaf. *Integrative Plant Biology*, 47 (10): 1228–1235, 2005b. DOI: 10.1111/j.1744-7909.2005.00154.x Cited on page(s) 112, 114

Y. Zhao, L. Zhang, P. Li, and B. Huang. Classification of high spatial resolution imagery using improved Gaussian Markov random-field-based texture features. *IEEE Trans. Geosci. Rem. Sens.*, 45 (5): 1458–1468, May 2007. DOI: 10.1109/TGRS.2007.892602 Cited on page(s) 50, 51

Q. Zhu and L.M. Collins. Application of feature extraction methods for landmine detection using the Wichmann/Niitek ground-penetrating radar. *IEEE Trans. Geosc. Rem. Sens.*, 43 (1): 81–85, Jan 2005. DOI: 10.1109/TGRS.2004.839431 Cited on page(s) 74

J. Zhuang and X. Xu. Genetic algorithms and its application to the retrieval of component temperature. *Remote Sens. Land Resour.*, 1: 28–33, 2000. Cited on page(s) 113

B. Zhukov, D. Oertel, F. Lanzl, and G. Reinhackel. Unmixing-based multisensor multiresolution image fusion. *IEEE Trans. Geosc. Rem. Sens.*, 37 (3): 1212–1226, May 1999. DOI: 10.1109/36.763276 Cited on page(s) 86

R. Zurita-Milla, L. Gómez Chova, L. Guanter, J. G. P. W. Clevers, and G. Camps Valls. Multitemporal unmixing of medium-spatial-resolution satellite images: A case study using MERIS images for land-cover mapping. *IEEE Trans. Geosc. Rem. Sens.*, 46 (4), 2011. Cited on page(s) 83, 85

Author Biographies

GUSTAVO CAMPS-VALLS

Gustavo Camps-Valls received a Ph.D. degree in Physics from the Universitat de València, Spain in 2002. He is currently an Associate Professor in the Department of Electronics Engineering and leading researcher at the Image Processing Laboratory (IPL) in the same university. Recently included in the ISI lists as a highly cited researcher, he co-edited the books "Kernel methods in bioengineering, signal and image processing" (IGI, 2007) and "Kernel methods for remote sensing data analysis" (Wiley & sons, 2009). Dr. Camps-Valls serves on the Program Committees of International Society for Optical Engineers (SPIE) Europe, International Geoscience and Remote Sensing Symposium (IGARSS), Machine Learning for Signal Processing (MLSP), and International Conference on Image Processing (ICIP). Since 2007 he is member of the Data Fusion technical committee of the IEEE Geoscience and Remote Sensing Society, and since 2009 he is member of the Machine Learning for Signal Processing Technical Committee of the IEEE Signal Processing Society. He is involved in the EUMETSAT MTG-IRS Science Team. He is Associate Editor of the "ISRN Signal Processing Journal" and "IEEE Geoscience and Remote Sensing Letters". Visit http://www.uv.es/gcamps for more information.

DEVIS TUIA

Devis Tuia was born in Mendrisio, Switzerland, in 1980. He received a diploma in Geography at the University of Lausanne in 2004, a Master of Advanced Studies in Environmental Engineering at the Federal Institute of Technology of Lausanne (EPFL) in 2005 and a Ph.D. in environmental sciences at the University of Lausanne in 2009. He was a postdoc researcher at both the University of València, Spain and the University of Colorado at Boulder under a Swiss National Foundation program. He is now a research associate at the Laboratoire des Systèmes d'Information Géographiques, EPFL. His research interests include the development of algorithms for information extraction and classification of high resolution remote sensing images and socio-economic data using machine learning algorithms, with particular focus in human-machine interaction and adaptation problems. Visit `http://devis.tuia.googlepages.com/` for more information.

LUIS GÓMEZ-CHOVA

Luis Gómez-Chova received the B.Sc. (with first-class honors), M.Sc., and Ph.D. degrees in electronics engineering from the University of Valencia, Valencia, Spain, in 2000, 2002 and 2008, respectively. He was awarded by the Spanish Ministry of Education with the National Award for Electronics Engineering. Since 2000, he has been with the Department of Electronics Engineering, University of Valencia, first enjoying a research scholarship from the Spanish Ministry of Education and currently as an Assistant Professor. He is also a researcher at the Image Processing Laboratory (IPL), where his work is mainly related to pattern recognition and machine learning applied to remote sensing multispectral images and cloud screening. He conducts and supervises research on these topics within the frameworks of several national and international projects. He is the author (or coauthor) of 30 international journal papers, more than 90 international conference papers, and several international book chapters. Visit `http://www.uv.es/chovago` for more information.

SANDRA JIMÉNEZ

Sandra Jiménez received a MSc in Mathematics and Computer Science in 2010. She's now an Associate Researcher in the Image Processing Laboratory (IPL) at the Universitat de València, where she is currently pursuing her PhD degree on image statistics and their applications in theoretical neuroscience and image analysis.

JESÚS MALO

Jesús Malo was born in 1970 and received the M.Sc. degree in Physics in 1995 and the PhD. degree in Physics in 1999, both from the Universitat de València. He was the recipient of the Vistakon European Research Award in 1994. In 2000 and 2001 he worked as Fulbright Postdoc at the Vision Group of the NASA Ames Research Center (with A.B. Watson), and at the Lab of Computational Vision of the Center for Neural Science, New York University (with E.P. Simoncelli). Currently, he is a leading researcher at the Image Processing Laboratory (IPL) at the Universitat de València. He is member of the Asociación de Mujeres Investigadoras y Tecnólogas (AMIT), and Associate Editor of IEEE Transactions on Image Processing. He is interested in models of low-level human vision, their relations with information theory, and their applications to image processing and vision science experimentation. His interests also include (but are not limited to) Fourier, Matlab, Equipo Crónica, Jim Jarmusch, Jordi Savall, Lou Reed, Belle and Sebastian, The Pixies, Milo Manara, la Bola de Cristal, Faemino y Cansado, and beauty in general.

Index